Efficient Screening Techniques to Identify Mutants with TR4 Resistance in Banana

Joanna Jankowicz-Cieslak • Ivan L. Ingelbrecht
Editors

Efficient Screening Techniques to Identify Mutants with TR4 Resistance in Banana

Protocols

IAEA

Joint FAO/IAEA Centre
Nuclear Techniques in Food and Agriculture

Springer

Editors
Joanna Jankowicz-Cieslak
Plant Breeding and Genetics Laboratory,
Joint FAO/IAEA Centre of Nuclear
Techniques in Food and Agriculture,
IAEA Laboratories Seibersdorf,
International Atomic Energy Agency
Vienna International Centre
Vienna, Austria

Ivan L. Ingelbrecht
Plant Breeding and Genetics Laboratory,
Joint FAO/IAEA Centre of Nuclear Techniques
in Food and Agriculture,
IAEA Laboratories Seibersdorf,
International Atomic Energy Agency
Vienna International Centre
Vienna, Austria

ISBN 978-3-662-64917-6   ISBN 978-3-662-64915-2   (eBook)
https://doi.org/10.1007/978-3-662-64915-2

Preface

Globally, bananas are the fourth most important food crop and are consumed both as a fruit and as a staple. Around 85% of the global production is destined for local markets in Southern countries while 15% enters the international trade. As such, bananas play a major role as a source of food, income and employment for smallholder farming systems in the South, in addition to being an important internationally traded commodity.

Currently, a deadly fungal disease, Fusarium wilt, is a major threat to banana production in all parts of the world. Fusarium wilt (FW) is caused by the soil-borne fungus *Fusarium oxysporum* f. sp. *cubense* (Foc). The fungus enters the banana plant through the roots and travels into the trunk and leaves where it causes premature wilting and eventually kills the plant. Since the nineteenth century, several strains of Foc have emerged. In the 1990s, a new strain of FW appeared in Asia that devastated banana plantations that were hitherto resistant to FW. This new strain is referred to as 'Tropical Race 4' or TR4. From Asia, TR4 spread to Australia, Africa and, very recently, also to Latin America, the most important banana exporting region globally. Foc TR4 is the most devastating of all FW strains because it not only affects Cavendish bananas but also many other bananas grown by small-scale farmers in Africa, Asia and Latin America. Therefore, Fusarium wilt TR4 has become a matter of international importance not only for the banana export industry but also for income generation and food security in smallholder farming systems in Southern countries.

Host plant resistance is a fundamental component for integrated management of Fusarium wilt in banana. Conventional cross-breeding of banana is hindered by several constraints, including polyploidy and low reproductive fertility in nearly all domesticated bananas. Mutagenesis techniques using physical or chemical mutagens offer an attractive, alternative approach to generate novel genetic diversity for banana genetic improvement, given that these methods do not require seed for the mutagenesis treatment.

To help address the banana Fusarium wilt TR4 pandemic, the FAO/IAEA Plant Breeding and Genetics subprogram launched a Coordinated Research Project (CRP) titled 'Efficient Screening Techniques to Identify Mutants with Disease Resistance for Coffee and Banana'. The CRP brought together leading banana breeders and experts from across the world who used mutagenesis techniques in combination with innovative biotechnology tools to develop and screen mutant populations for resistance to Fusarium wilt TR4.

This book comprises a collection of protocols ensuing from their efforts. The protocol chapters cover conventional and innovative methods for the preparation and mutagenesis of target explants or cells for mutation induction together with lab-, greenhouse- and field-based screening techniques specifically for Fusarium wilt. Low-cost methodologies for *in vitro* multiplication are also described as well as advanced genomics and bioinformatics tools for mutation discovery. While the CRP and protocols were mainly focused on Cavendish bananas, similar technologies can be adapted to other banana varieties, including cooking banana, an important staple food in tropical countries.

The main purpose of this protocol book is to widely disseminate the methods and techniques developed under this CRP with FAO/IAEA Member States involved in banana breeding generally and especially those countries faced with the new threat of Fusarium wilt TR4.

Plant Breeding and Genetics Laboratory,   Ivan L. Ingelbrecht
Joint FAO/IAEA Centre of Nuclear
Techniques in Food and Agriculture,
IAEA Laboratories Seibersdorf,
International Atomic Energy Agency,
Vienna International Centre,
Vienna, Austria

Acknowledgements

We would like to thank the entire Plant Breeding and Genetics team, past and present, especially Dr Bradley J. Till and Dr Stephan Nielen, involved in the development and implementation of the IAEA Coordinated Research Project (CRP) "Efficient Screening Techniques to Identify Mutants with Disease Resistance for Coffee and Banana" (code: D22005). Special thanks to the CRP participants for sharing their unique expertise in banana breeding for Fusarium wilt TR4 resistance and contributing the protocols. We further acknowledge the reviewers of the book chapters for their valuable inputs. Funding for this CRP and supporting research at the PBG Laboratory work was provided by the Food and Agriculture Organization of the United Nations and the International Atomic Energy Agency through their Joint FAO/IAEA Centre of Nuclear Techniques in Food and Agriculture, with additional support from Belgium through the Peaceful Uses Initiative "Enhancing climate change adaptation and disease resilience in banana-coffee cropping systems in East Africa" (code: EBR-BEL01-18-03) and the USA through the support of an associate research officer (plant pathology) (code: EBR-USA07-13-13).

Contents

Contributors

Faiz Ahmad Agrotechnology and Biosciences Division, Malaysian Nuclear Agency, Ministry of Science, Technology and Innovation Malaysia (MOSTI), Kajang, Selangor, Malaysia

Zaiton Ahmad Agrotechnology and Biosciences Division, Malaysian Nuclear Agency, Ministry of Science, Technology and Innovation Malaysia (MOSTI), Kajang, Selangor, Malaysia

Mustapha Akil Agrotechnology and Biosciences Division, Malaysian Nuclear Agency, Ministry of Science, Technology and Innovation Malaysia (MOSTI), Kajang, Selangor, Malaysia

Sakinah Ariffin Agrotechnology and Biosciences Division, Malaysian Nuclear Agency, Ministry of Science, Technology and Innovation Malaysia (MOSTI), Kajang, Selangor, Malaysia

Sheryl Bothma Department of Plant Pathology, Private Bag X1, Stellenbosch University, Matieland, South Africa

Catherine Breton Alliance Bioversity International-CIAT, Parc Scientifique Agropolis II, Montpellier, France

Alberto Cenci Alliance Bioversity International-CIAT, Parc Scientifique Agropolis II, Montpellier, France

Rachel Chase Alliance Bioversity International-CIAT, Parc Scientifique Agropolis II, Montpellier, France

Babita Jhurree-Dussoruth Food and Agricultural Research and Extension Institute, Reduit, Mauritius

Florian Goessnitzer Plant Breeding and Genetics Laboratory, Joint FAO/IAEA Centre of Nuclear Techniques in Food and Agriculture, IAEA Laboratories Seibersdorf, International Atomic Energy Agency, Vienna International Centre, Vienna, Austria

Affrida Abu Hassan Agrotechnology and Biosciences Division, Malaysian Nuclear Agency, Ministry of Science, Technology and Innovation Malaysia (MOSTI), Kajang, Selangor, Malaysia

Chunhua Hu Institute of Fruit Tree Research, Guangdong Academy of Agricultural Sciences; Key Laboratory of South Subtropical Fruit Biology and Genetic Resource Utilization; Ministry of Agriculture and Rural Affairs, Guangdong Province Key Laboratory of Tropical and Subtropical Fruit Tree Research, Guangzhou, China

Rusli Ibrahim Agrotechnology and Biosciences Division, Malaysian Nuclear Agency, Ministry of Science, Technology and Innovation Malaysia (MOSTI), Kajang, Selangor, Malaysia

Ivan L. Ingelbrecht Plant Breeding and Genetics Laboratory, Joint FAO/IAEA Centre of Nuclear Techniques in Food and Agriculture, IAEA Laboratories Seibersdorf, International Atomic Energy Agency, Vienna International Centre, Vienna, Austria

Joanna Jankowicz-Cieslak Plant Breeding and Genetics Laboratory, Joint FAO/IAEA Centre of Nuclear Techniques in Food and Agriculture, IAEA Laboratories Seibersdorf, International Atomic Energy Agency, Vienna International Centre, Vienna, Austria

Behnam Naserian Khiabani Plant Breeding Department, Nuclear Agriculture Research School, Nuclear Science and Technology Research Institute (NSTRI), Tehran, Iran

Gaudencia A. Lantican Dole Philippines, Inc., Davao City, Philippines

Anis Nadia Mohd Faisol Mahadevan Agrotechnology and Biosciences Division, Malaysian Nuclear Agency, Ministry of Science, Technology and Innovation Malaysia (MOSTI), Kajang, Selangor, Malaysia

George Mahuku International Institute of Tropical Agriculture (IITA) Regional Hub, Dar-es-Salaam, Tanzania

Diane Mostert Department of Plant Pathology, Private Bag X1, Stellenbosch University, Matieland, South Africa

Privat Ndayihanzamaso Department of Plant Pathology, Private Bag X1, Stellenbosch University, Matieland, South Africa

Norazlina Noordin Agrotechnology and Biosciences Division, Malaysian Nuclear Agency, Ministry of Science, Technology and Innovation Malaysia (MOSTI), Kajang, Selangor, Malaysia

Mathieu Rouard Alliance Bioversity International-CIAT, Parc Scientifique Agropolis II, Montpellier, France

Max Ruas Alliance Bioversity International-CIAT, Parc Scientifique Agropolis II, Montpellier, France

Nicolas Roux Alliance Bioversity International-CIAT, Parc Scientifique Agropolis II, Montpellier, France

Julie Sardos Alliance Bioversity International-CIAT, Parc Scientifique Agropolis II, Montpellier, France

Bradley J. Till Veterinary Genetics Laboratory, University of California, Davis, Davis, CA, USA

Altus Viljoen Department of Plant Pathology, Private Bag X1, Stellenbosch University, Matieland, South Africa

Yuanli Wu Institute of Fruit Tree Research, Guangdong Academy of Agricultural Sciences; Key Laboratory of South Subtropical Fruit Biology and Genetic Resource Utilization; Ministry of Agriculture and Rural Affairs, Guangdong Province Key Laboratory of Tropical and Subtropical Fruit Tree Research, Guangzhou, China

Ganjun Yi Institute of Fruit Tree Research, Guangdong Academy of Agricultural Sciences; Key Laboratory of South Subtropical Fruit Biology and Genetic Resource Utilization; Ministry of Agriculture and Rural Affairs, Guangdong Province Key Laboratory of Tropical and Subtropical Fruit Tree Research, Guangzhou, China

Chapter Reviewers

Souleymane Bado Plant Science (PHYTOPRISE GmbH), PHYTONIQ GmbH, Oberwart, Austria

Nadiya Akmal Baharum Faculty of Biotechnology and Biomolecular Sciences, Department of Cell and Molecular Biology, Universiti Putra Malaysia, Serdang, Selangor, Malaysia

Ratri Boonruangrod Department of Horticulture, Faculty of Agriculture at KPS, Kasetsart University, Nakhon Pathom, Thailand

Azam Borzouei Agricultural Research School, Nuclear Science and Technology Research Institute, Karaj, Iran

Christopher Cullis Department of Biology, Case Western Reserve University, Cleveland, Ohio, USA

Jessada Doungkeow Department of Horticulture, Faculty of Agriculture at KPS, Kasetsart University, Nakhon Pathom, Thailand

Ilona Czyczylo-Mysza The Franciszek Górski Institute of Plant Physiology Polish Academy of Sciences, Krakow, Poland

Alexandra zum Felde Independent Consultant, Witzenhausen, Germany

Stanley Freeman Department of Plant Pathology and Weed Research, ARO, The Volcani Center, Rishon LeZion, Israel

Prateek Gupta Department of Genetics, Hebrew University of Jerusalem, Jerusalem, Israel

Sobri Hussein Agrotechnology and Biosciences, Malaysian Nuclear Agency, Kuala Lumpur, Malaysia

Rusli Ibrahim Agrotechnology and Biosciences, Malaysian Nuclear Agency, Kuala Lumpur, Malaysia

Praphat Kawicha Kasetsart University, Sakon Nakhon, Thailand

Part I
Mutation Induction

Chapter 1
Induced Mutagenesis and *In Vitro* Mutant Population Development in *Musa spp.*

Joanna Jankowicz-Cieslak, Florian Goessnitzer, Bradley J. Till, and Ivan L. Ingelbrecht

Abstract Mutagenesis of *in vitro* propagated bananas is an efficient method to introduce novel alleles and broaden genetic diversity. Mutations can be induced by treatment of plant cells with chemical mutagens or ionizing radiation. The FAO/IAEA Plant Breeding and Genetics Laboratory established efficient methods for mutation induction of *in vitro* shoot tips in banana using physical and chemical mutagens as well as methods for the efficient discovery of EMS-induced single nucleotide mutations in targeted genes or amplicons and identification of large genomic changes, including deletions and insertions. Mutagenesis of *in vitro* propagated tissues requires large populations serving as starting material, and a long process to dissolve genetic mosaics (chimeras) resulting from the mutagenesis of multicellular tissues. However, treating shoot apical meristems of tissue cultured bananas with a mutagen is a commonly used practice for banana mutation breeding programmes, and still the most effective. In our previous studies, we showed that chimeras, unique mutations accumulated in different cells of the plant propagule, could be rapidly removed *via* isolation of shoot apical meristems and subsequent longitudinal bisection. Further, induced mutations were maintained in mutant plants for several generations. We established such systems for inducing and maintaining both point mutations caused *via* EMS mutagenesis as well as insertions and deletions caused by gamma irradiation and describe hereafter methods for dose selection, gamma irradiation and chimera dissolution.

Keywords Mutation induction · Gamma mutagenesis · Banana · Polyploidy · Fusarium TR4

J. Jankowicz-Cieslak (✉) · F. Goessnitzer · I. L. Ingelbrecht
Plant Breeding and Genetics Laboratory, Joint FAO/IAEA Centre of Nuclear Techniques in Food and Agriculture, IAEA Laboratories Seibersdorf, International Atomic Energy Agency, Vienna International Centre, Vienna, Austria
e-mail: j.jankowicz@iaea.org

B. J. Till
Veterinary Genetics Laboratory, University of California, Davis, CA, USA

© The Author(s) 2022
J. Jankowicz-Cieslak, I. L. Ingelbrecht (eds.), *Efficient Screening Techniques to Identify Mutants with TR4 Resistance in Banana*,
https://doi.org/10.1007/978-3-662-64915-2_1

1 Introduction

Bananas and plantains are among the most important staple food crops for people living in tropical and subtropical countries. They are herbaceous monocots belonging to the genus *Musa*; most are seedless, polyploid, sterile and clonally propagated. The majority of banana and plantains are consumed locally.

Vegetatively propagated crops (VPCs) such as banana pose unique problems compared to cereals because they have a reduced genetic diversity as they can't be cross pollinated to enhance variation. Furthermore, because of its triploid parthenocarpic nature, bananas do not produce seeds and conventional breeding is thus a long process. Advances in biotechnology for crop improvement have had a great impact on vegetatively propagated crops (Gosal et al. 2010). Biotechnology based on tissue culture is complementary to conventional breeding technology.

One advantage of vegetatively propagated crops is that methods have been developed for rapid clonal propagation. Micropropagation is currently used in many countries for rapid propagation of disease-free planting material for distribution on a large scale. Such tissue culture techniques that ensure genetic stability (e.g. using shoot tip/nodal cultures for propagation) are particularly useful for *in vitro* mutation induction and maintenance of mutant plant populations. This technique is also of particular interest to breeders since the multiplication of the new lines for field trials and evaluation could be hastened, thereby shortening the time required for the release of new cultivars. *In vitro* techniques also offer possibilities to use induced mutation for further manipulation aimed at improvement. New genetic variation conferring a desirable trait can be fixed, and identical material rapidly deployed.

The structure of meristematic regions and the development of new meristems from differentiated tissue are particularly important when investigating radiation-induced mutation of vegetatively propagated crops (VPCs). In most cases, the new shoots originate from a single epidermal cell from a tissue, and this could directly lead to homohistant mutant plants whose genetics may be investigated further (Spencer-Lopes et al. 2018).

One of most critical prerequisites for successful mutation breeding is the determination of the optimal mutagen dose. The dose required for a particular experiment depends on the desired effects but may be restricted by undesirable effects of the mutagenic treatment, which could lead to lethality. There is a strong correlation between the genotype and the sensitivity of the plant material to the mutagenic treatments in plants (Jankowicz-Cieslak et al. 2012). The dose increase causes severe mutations, such as chromosomal aberrations, and can cause cell damage and subsequently death. While little is known for VPCs, the data from seed propagated plants suggests that fine-tuning of dose applied may be needed. Therefore, radiation sensitivity tests should be carried out to determine the mutagen dose that results in a 50 percent reduction in e.g., plant height or plant weight. In practice, a breeder applying irradiation treatment on vegetatively propagated crops may decide to settle for a growth reduction of 30–50 percent (GR30–50) for M_1V_1 plants or a survival rate of 40–60 percent depending on the sensitivity of the plant material. An equal number of control materials for the comparison should be planted at the same time.

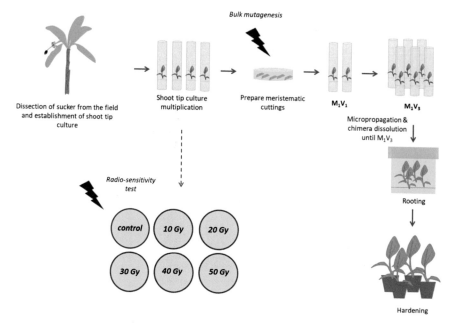

Fig. 1.1 Mutation induction pipeline using banana *in vitro* shoot tip cultures. The first step of the process consists of establishment of banana cultures, either *via* sucker dissection from the field or alternatively obtaining accessions of interest from germplasm repository. Upon sufficient multiplication of banana shoot tips, the bulk mutagenesis can be performed or, if the knowledge on radiosensitivities of this particular genotype doesn't exist, a radiosensitivity test should be carried out. After mutagenic treatment of a bulk number of shoot tips, chimeras need to be dissolved and at the M_1V_3 stage plantlets can be rooted and hardened for field or screenhouse selection

Measurement of growth reduction of *in vitro* treated material should be recorded 30 days after the treatments.

For VPCs, various methods have been developed which involve tissue isolation and dissection during post-mutagenesis aiming at reducing the genotypic complexity of the resulting plants.

The following procedures are routinely used in the Plant Breeding and Genetics Laboratory to micropropagate banana and plantain by shoot tip culture and perform mutagenesis (Fig. 1.1).

2 Materials

2.1 **In Vitro** Culture Media

1. Stock solutions prepared under sterile conditions and vacuum filtering after chemicals are completely dissolved in the given amount of distilled water.
   (a) Thiamine hydrochloride Stock: 100 mg   in 100 ml dest. Water

(b) 6-benzylaminopurine (BAP):   112.6 mg in 0.5 l dest. Water
(c) L-Cysteine HCl: 2 g   in 0.5 l dest. Water
(d) Indole-3-butyric acid (IBA): 20.3 mg   in 100 ml dest. Water
(e) Isopentenyl adenine (2iP):   20.32 mg in 100 ml dest. Water

2. Initiation medium / Maintenance medium composition (1 l).

(a)	Thiamine hydrochloride	1 ml
(b)	6-benzylaminopurine (BAP)	20 ml
(c)	L-Cysteine HCl	10 ml
(d)	Sucrose	40 g

3. Rooting medium composition (1 l).

(a)	B5 Vitamin	1 ml
(b)	2iP	5 ml
(c)	IBA	0.1 ml
(d)	L-Cysteine HCl	1 ml
(e)	Sucrose	40 g

4. Sucrose (household grade).
5. Murashige and Skoog basal salt with minimal organics (MS) (*e.g.* Sigma Cat Nr.: M-6899).
6. Tissue culture grade water.
7. Gelling agent (*e.g.* Gelrite).
8. 0.22 μm Cellulose Acetate (CA) sterilising, low binding filters.
9. Vacuum pump.
10. Sterile tubes (50 ml).
11. Analytical balance.
12. Weighing trays.
13. Spatula.
14. Magnetic stir bar.
15. Hot plate.
16. pH meter.
17. NaOH.
18. KOH.
19. Medium dispenser.
20. Culture tubes.
21. Closures for culture tubes.
22. Erlenmeyer flasks (250 ml).
23. Aluminium foil.
24. Autoclave.
25. Cold room for media storage.

2.2 Gamma-Ray Mutagenesis

1. Physical mutagenesis source.
2. Tissue culture laboratory equipped with flow benches (fitted with gas).
3. Sterile S-27 liquid culture medium.
4. High quality, disease-free *in vitro* cuttings (*e.g.* 1000 propagules per treatment).
5. Petri plates (94 mm and 145 mm diameter).
6. Whatman filter papers for 90 mm Petri dishes.
7. Sterile distilled water.
8. Sterile forceps.
9. Sterile scalpel.
10. Parafilm.
11. Orbital shaker.
12. Growth rooms with light and temperature control (light regime 65 µmol/m^2/s; *e.g.* Cool White fluorescent tubes, Philips TLP 36/86; temperature regime of 22° +/− 2 °C).
13. Personal protective equipment (laboratory coat, shoe protection, nitril gloves).

2.3 Calculation of Lethality and Growth Reduction

1. Tissue culture laboratory equipped with sterile flow benches (fitted with gas).
2. Analytical balance.
3. Sterile Petri plates for weighing *in vitro* material (94 mm).
4. Sterile Whatman filter papers for 90 mm Petri dishes.
5. Forceps.
6. Sterile S-27 liquid culture medium.
7. Standard spreadsheet software *e.g.* Microsoft Excel.

2.4 Chimera Dissolution

1. Tissue culture laboratory equipped with flow benches (fitted with gas).
2. Growth rooms with light and temperature control (light regime 65 µmol/m^2/s; *e.g.* Cool White fluorescent tubes, Philips TLP 36/86; temperature regime of 22° +/− 1 °C).
3. Sterile S-27 liquid culture medium.
4. Sterile S-27 solid culture medium.
5. Ethanol for surface and tools sterilisation.
6. Forceps.
7. Scalpels.

8. Scalpel blades.
9. Petri plates (94 mm and 145 mm diameter).
10. Orbital shaker.

2.5   Hardening

1. Facility with light, temperature and humidity control or plastic bags to cover pots.
2. Peat.
3. Sand.
4. Soil mixer (for large amounts).
5. Running tap water.
6. 5 l Container with room temperature tap water.
7. Small trays or jiffy pots.
8. Forceps.
9. Plastic labels.
10. Permanent pen or pencil.

3   Methods

3.1   Preparation of Liquid Maintenance Medium S-27

1. Prepare stock solutions of thiamine (1 mg/ml), BAP (0.23 mg/ml) and L-Cysteine (4 g/l).
2. Filter sterilise stock solutions.
3. Dispense into 50 ml batches and freeze for future use. The working solution should be stored at 4 °C.
4. For 1 litre of the liquid culture media, use the following: 4.4 g of Murashige and Skoog basal salt with minimal organics, 40 g sucrose, 10 ml L-Cysteine, 20 ml BAP and 1 ml of thiamine stock solutions. Use double distilled water.
5. Place the media on the mixer and let it mix for 30 min.
6. Calibrate the pH meter as per manufacturer instruction.
7. While stirring, adjust medium to pH 5.8 using NaOH and HCl.
8. Dispense 12 ml of the culture medium per Erlenmeyer flask.
9. Close each Erlenmeyer flask tightly over the top with an aluminium foil.
10. Autoclave for 20 min at 121 °C at 1 bar.
11. Allow the medium to cool to room temperature.
12. Store the medium in a cold room (*see* Note 1).

3.2 Preparation of Solid Culture Medium

1. For 1 litre of solid medium cook 1.8 g of Gelrite and 40 g sucrose in 400 ml of tissue culture grade water.
2. In a separate beaker containing 400 ml water mix:
   - (a) For initiation/maintenance medium: 4.4 g of Murashige and Skoog basal salt with minimal organics, 10 ml L-Cysteine, 20 ml BAP and 1 ml of thiamine stock solutions.
   - (b) For rooting medium: 4.4 g of Murashige and Skoog basal salt with minimal organics, 1 ml B5 Vitamin, 5 ml 2iP, 0.1 mL IBA and 1 ml of L-Cysteine HCl stock solution.
3. Heat the mixture while stirring.
4. When completely dissolved and hot enough add the Gelrite/sucrose mixture.
5. Heat while stirring until the solution is homogenous and clear, add tissue culture grade water to 1 litre.
6. While stirring adjust the pH to 5.8 using NaOH and HCl.
7. Dispense 8 ml of the medium into culture tubes.
8. Sterilise the medium for 20 min at 121 °C.
9. Allow medium to cool to room temperature prior to use.

3.3 Initiation and Maintenance of Banana Cultures

1. Select disease free vegetative buds or peepers or juvenile sword suckers (*see* Note 2).
2. Cut the explants to ca. of 5 mm in size consisting of the meristematic dome with two to four leaf primordia.
3. Wash explants under running tap water to remove dirt and keep in covered Petri dishes. From this point on all procedures need to be carried out in the sterile airflow cabinet.
4. Rinse explants with 70% ethanol for surface sterilization.
5. Transfer them in 40% commercial Chlorox with a drop of tween 20 (any available brand of bleach in a final concentration of 0.5–1% NaOCI can be used) for 20 min.
6. Agitate explants several times while in Chlorox solution.
7. Wash the explants with sterilized water three times.
8. Remove two or three leave sheaths, not damaging the meristem, and transfer to a solid initiation medium.
9. Cultures are kept at 28 °C for 16 h light/day (65 µmol/m^2/s; Cool White fluorescent tubes, Philips TLP 36/86).
10. Monitor cultures for contamination on a daily basis (*see* Note 3).
11. Once plantlets are fully developed, increase the number of plantlets by micropropagation of shoot tip meristems.

12. Isolate meristematic tips from each plantlet and propagate through longitudinal division into two propagules using a scalpel.
13. Transfer isolated meristematic shoot tips into culture flasks containing fresh liquid media.
14. Place the flasks on a horizontal gyratory shaker at 60 rpm and allow the explants to grow for approx. 30 days (light 65 µmol/m²/s; Cool White fluorescent tubes, Philips TLP 36/86 and the temperature of 26° +/− 1 °C).
15. Repeat the process in order to increase the population size for mutagenesis purposes (*see* Note 4).

3.4 *Establishment of the Radiosensitivity Curve and Calculation of Growth Reduction (GR)*

3.4.1 Tissue Irradiation

1. Autoclave all non-disposable materials (*e.g.* sieves, forceps).
2. Prepare fresh liquid S-27 culture medium and autoclave.
3. Prepare meristematic cuttings per each treatment chosen (*see* Note 5).
4. Place cuttings into a Petri plate with a drop of water or on a moist filter paper (*see* Note 6).
5. Calculate the exposure time (sec., min.) based on the dose rate of gamma cell irradiator that will be used (Gy/s or Gy/min) (*see* Note 7).
6. Label Petri plates according to the dose required and genotype when handling more than one genotype. Include an untreated Petri plate as control.
7. Wrap all Petri plates with Parafilm in order to avoid surface contamination.
8. Transfer the material for irradiation to the physical mutagen source laboratory (*see* Note 8).
9. Apply the required dose (Table 1.1) by placing Petri plates into irradiator chamber for the exposure time to produce the dosage (*see* Note 9).
10. Irradiate each Petri plate with the chosen dose.
11. Transfer irradiated samples back to the tissue culture laboratory.
12. Surface sterilize each Petri plate with 70% ethanol before placing them in the laminar flow bench.
13. Remove the Parafilm under sterile environment.

Table 1.1 Example of doses chosen for mutagenic treatment of *in vitro* banana

Target dose* (Gy)	Time (sec)	Min	Sec	Calculated dose (Gy)
10	6	0	6	11.16
20	15	0	15	21.21
30	24	0	24	31.26
40	33	0	33	41.30
50	42	0	42	51.35

14. Transfer the irradiated explants (5 explants per flask) to sterile, labelled conical flasks (Erlenmeyer flasks) containing 12 ml liquid S-27 medium (*see* Note 10).
15. With the non-irradiated control sample, place the flasks on a horizontal gyratory shaker at 60 rpm (light 65 μmol/m^2/s; Cool White fluorescent tubes, Philips TLP 36/86 and the temperature of 26° +/− 1 °C).
16. If necessary, transfer cultures weekly into fresh liquid media to reduce possible accumulation of phenolic compounds due to the stress caused by mutagenesis (*see* Note 11).
17. Thirty days after the treatment, take measurements of the chosen parameters (*e.g.* weight of the explants) and record survival rates of the mutagenized population (*see* Note 12).

3.4.2   Data Collection

1. After 30 days of incubation, assess the viability of plants by counting the surviving plantlets and measuring the fresh weight of each plant (*see* Note 13 and Fig. 1.2).
2. Place a balance in the laminar bench and weigh each mutated plant separately in a sterile Petri plate.
3. Record the data for each treatment and enter it into a spreadsheet (*e.g.* Microsoft Excel) (*see* Note 14).

3.4.3   Data Analysis

1. For every parameter measured, calculate the average per treatment for each replication.
2. Repeat point 1 for control material.
3. Express the reduction/increase of the radiated materials as a percentage of the non-irradiated control (*see* Note 15).
4. Plot the calculated values as a dose-response curve using the dose as x-axis and the percentages on the y-axis. The control treatment is set as 100 percent (*see* Fig. 1.2).
5. Repeat these steps with all the measured parameters (reduction in fresh weight percentage, plantlet height, root length, survival) (*see* Note 16).
6. Compare the graphs from the different parameters (*see* Note 17).
7. Estimate the mutagen dose required to obtain e.g. 50% of the control (*see* Fig. 1.2, red line).
8. Identify the doses suitable for bulk mutagenesis of your material (*see* Note 18 and Table 1.2).

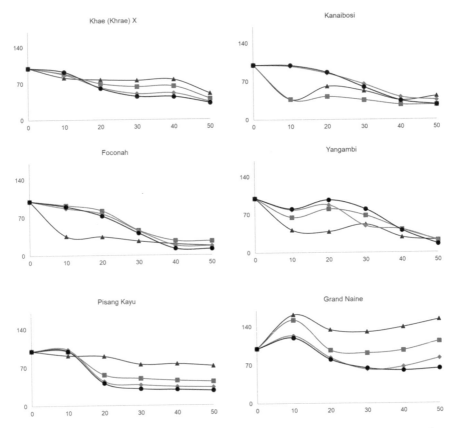

Fig. 1.2 Effect of irradiation on propagules weight of six different banana accessions expressed as percentage of non-irradiated (control) and percentage reduction in mean weight. Blue, orange, green and red lines indicate measurements taken with weekly intervals (1 week – blue, 2 weeks – orange, 3 weeks – green and 4 weeks – red post mutagenesis). A clear trend in response stabilisation takes place already after 2 (orange) to 3 (green) weeks

Table 1.2 Examples of banana sensitivities expressed as GR30 and GR50 values calculated from radiosensitivity graphs (Fig. 1.2 and *see* Note 19)

Accession Name	Accession number	Species / Group	SubSpecies / SubGroup	GR30 (Gy)	GR50 (Gy)
Khae (Khrae) X	ITC0660	Acuminata	Subsp. siamea	17	25
Kanaibosi				26	33
Foconah	ITC0649	AAB	Subgr. Pome	21	28
Yangambi	ITC1123	AAA	Subgr. Ibota	33	38
Pisang Kayu	ITC0420	AAA	Subgr. Orotava	15	18
Grand Naine	ITC0180	AAA	Subgr. Cavendish	25	NA

3.5 Bulk Mutagenesis by Physical Agents

1. Autoclave all non-disposable materials (*e.g.* sieves, forceps).
2. Prepare fresh liquid S-27 culture medium and autoclave.
3. Prepare meristematic cuttings (*e.g.* 1000) per treatment chosen (*see* Note 20).
4. Distribute cuttings evenly into Petri plates with a drop of water or on a moist filter paper.
5. Calculate the exposure time (sec., min.) based on the dose rate of the gamma cell irradiator that will be used (Gy/s or Gy/min).
6. Label Petri plates according to the dose required and genotype when handling more than one genotype. An untreated Petri plate (control) with the same number of propagules will be held in the same conditions as the treated ones.
7. Transfer the Petri dishes to physical mutagen source laboratory.
8. Apply the required dose by placing Petri plates into irradiator chamber for the exposure time to produce the dosage.
9. Transfer irradiated samples to the tissue culture laboratory.
10. Surface sterilize the Petri dish with 70% ethanol before removing the Parafilm.
11. Transfer the irradiated explants along with a non-irradiated control into sterile labelled conical flasks (Erlenmeyer flasks) containing 12 ml liquid S-27 medium (5 explants per flask).
12. Place flasks on a horizontal gyratory shaker at 60 rpm (light 65 μmol/m^2/s; Cool White fluorescent tubes, Philips TLP 36/86 and the temperature of 26° +/− 1 °C).
13. If necessary, transfer cultures weekly into fresh liquid media to reduce possible accumulation of phenolic compounds due to the stress caused by mutagenesis. Monitor cultures for the presence of any contamination.

3.6 Post-Mutagenesis Handling

3.6.1 Chimera Dissolution

1. Grow explants in liquid culture media under a constant horizontal rotation at 60 rpm with a continuous light (65 μmol/m^2/s; Cool White fluorescent tubes, Philips TLP 36/86) at 22° +/− 2 °C.
2. Each shoot tip is the starting material for subsequent cultures and is referred to as a line as previously described (Jankowicz-Cieslak et al. 2012; Jankowicz-Cieslak and Till 2016) (*see* Note 21).
3. In order to remove potential chimeric sectors, propagate shoot tip meristems through longitudinal division into two propagules using a scalpel (Fig. 1.3).
4. Define each mutated meristem as a source of an individual line and assign to it a number or a barcode, depending on the system you have (*see* Note 22). This number corresponds to the mutant line that will be generated from this individual meristem.

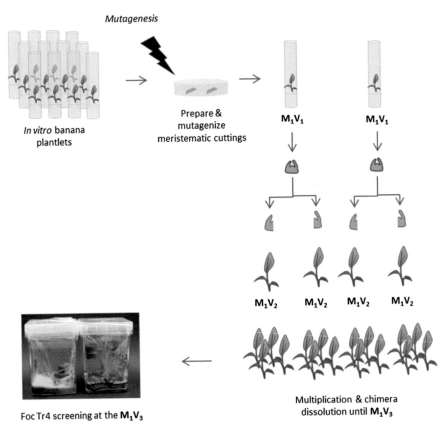

Fig. 1.3 Dissolution of chimeras in bananas. Plants after the mutagenic treatment are allowed to grow for 30 days in a liquid culture media at a constant rotary shaking (60 rpm). Each mutated plant is given a unique line number and assigned a population stage starting with M_1V_1. Chimeras may exist after 30 days recovery period if surviving meristematic cells harbour different mutations. In order to dissolve potential chimeras, meristems are isolated and divided into two parts through a longitudinal cut which results in most cases in generation of 2 daughter plants. These are allowed to recover and grow for another 30 days. The procedure is repeated 3 times. At the M_1V_3 stage, plants are being transferred into the solid culture media for long-term storage

5. Isolate meristematic tips from each of mutated plants.
6. Transfer every single isolated meristem into a separate culture tube or flask containing fresh liquid or solid media.
7. Incubate cultures for 4–5 weeks.
8. Repeat the process in order to make the population of M_1V_3 individuals (*see* Fig. 1.3 and *see* Note 23).
9. At each meristematic division transfer the material into fresh culture media. Remember to follow the nomenclature assigned for each line.

10. Transfer plants into the solid media for a long-term storage (culture media supplemented by 1.8% Gelrite) (*see* Note 24).
11. Maintain cultures at 22° +/− 2° under stationary incubation and 12 h light cycle for the duration of the study.

3.6.2 Shoot Elongation and Rooting

1. Transfer regenerants on to a rooting medium (containing half-strength MS, 1 mg/l thiamine, 5 μmol/l IBA, 20 g/l sucrose, and 7 g/l agar).
2. Maintain cultures at 22° +/− 2 °C under stationary incubation and 12 h light cycle.
3. After 3–5 weeks, transfer well-rooted plantlets to non-aseptic conditions or the greenhouse (*see* Note 25).

3.6.3 Hardening of *In Vitro* Mutant Population

1. Use facility with light, temperature and humidity control or plastic bags to cover pots with sufficient water in the soil to maintain high ambient humidity.
2. Prepare soil mixture of 50:50 (m/m) peat and sand by hand or soil mixture until both components are evenly mixed.
3. Fill pots with soil before starting the procedure.
4. Gently remove well rooted plantlets from culture vessels.
5. Wash remaining medium off through running room-temperature tap water.
6. Collect cleaned plantlets in a container with room-temperature water.
7. Transplant plantlets in previously prepared pots.
8. Water plantlets immediately after transfer to soil.
9. Keep surface of plants moistened until transferred to humidity chamber.
10. Decrease relative humidity weekly by 10% until 60% relative humidity in the chamber is reached (*see* Note 26).
11. After 3–4 weeks move plants to the greenhouse.
12. Allow plantlets to grow for about 4 weeks or until reaching sufficient size to be transferred to the greenhouse, nursery or field in preparation for screening and selection (*see* Note 27 and Chap. 7).

3.7 *Phenotypic and Genotypic Selection*

1. Screen mutagenized population using reverse or forward genetics methods (*see* Note 28 and Chaps. 5, 7, 8, 9).

4 Notes

1. It is recommended to prepare fresh media in the amount that is needed. If a cold room or refrigerator is available, media can be stored at 4–8 °C up to 6 weeks.
2. The mutation breeding process usually starts with *in vitro* propagation of the plant material using shoot tips, corms, and also embryonic cells suspensions. Shoot tips are found to be the most commonly used hence, the protocol described is for mutation induction of *in vitro* shoot tips.
3. In case of contamination, repeat the process of surface sterilisation.
4. Excessive blackening in some genotypes, (e.g., AAB, BB) is normally overcome by rinsing the explants in an antioxidant solution made from 50 mg/l citric acid, 50 mg/l ascorbic acid and 5 ml of commercial Chlorox.
5. Estimate how many explants can fit into a Petri plate. Usually, each treatment replicate should be placed in a separate Petri plate (*e.g.* 30 cuttings/plate), however if the number of explants is high, split them equally. When the applied dose for the genotype is unknown, a radiation test must be conducted. It is advisable that even if a dosage is known and was determined in a different laboratory, that a radiosensitivity test is performed as laboratory-to-laboratory differences can affect growth reduction measurements. To perform a radiosensitivity test on vegetative tissue, use 30 cuttings per dose with a wide range from 0 to 100 Gray (Gy). The range of 0, 10, 20, 30, 40, 50 and 60 Gy of gamma may be sufficient to establish the optimal dose due to the high moisture content of banana *in vitro* tissue in comparison to seeds. The same number of non-irradiated explants should be used as control.
6. The amount of water has an important impact on the irradiation efficiency. Be consistent with the water used for irradiation by using a device to transfer it (*e.g.* a pasteur pipette or micro pipette). In the case of a long-distance transfer of samples for mutagenesis/radiosensitivity, consider placing explants on a solid media (*e.g.* Petri plates, magenta boxes).
7. Exposure time is equal to the required dose divided by the dose rate of the day. The Gy is the unit used to quantify the absorbed dose of radiation (1 Gray = 1 J/kg).
8. Physical mutagens comprise all nuclear radiations and sources of radioactivity including ultraviolet light (a non-ionising radiation), several types of ionizing radiations, namely X- and gamma-rays, alpha and beta particles, protons and neutrons. This protocol describes gamma ray induced mutagenesis. Gamma ray mutagenesis may be performed using different facilities, such as gamma cell irradiator, gamma phytotron, gamma house, or gamma field (Spencer-Lopes et al. 2018). Gamma cells are the most commonly used emitters for plant mutation induction. The isotopes Cobalt-60 (^{60}Co) and Caesium-137 (^{137}Cs) are the main sources of gamma rays. The International Basic Safety Standards published by the IAEA provides details for the safe handling of these sources (IAEA 2014).

9. Radioactivity is mutagenic and carcinogenic. The safety precautions for exposing plant material to a gamma irradiation source must be strictly observed. Radioactive sources should therefore be operated only by trained and authorized personnel. Local regulations are usually explicit.

10. The gamma irradiated samples are safe to handle.

11. Blackening of liquid cultures can be the result of accumulation of phenolic compounds in the culture. Observe cultures for such media discoloration or for an eventual contamination. Remove contaminated flasks and decontaminate by autoclaving.

12. The dose of a mutagen that achieves the optimum mutation frequency with the least possible unintended damage, is regarded as the optimal dose for induced mutagenesis. For physical mutagens, this is estimated by carrying out tests of radiosensitivity. These estimates of the percentage in plant growth reduction are good parameters for estimating the damage due to mutagenic treatment. By reading off the dose corresponding to 50% reduction, the so-called lethal dose 50 (LD_{50}), is obtained. The LD_{50} is an appropriate dose for irradiation but, in practice, a range of doses around this value should be used. In case of VPCs and seed propagated plants, also a growth reduction (GR) is being taken into count. Additional tests can be carried out such as genome sequencing to identify the spectrum and density of induced mutations (*see* Chap. 8 and 9).

13. Previous studies indicated that the radiosensitivity reading can be taken only 30 days post-irradiation. Results presented in Fig. 1.2 clearly demonstrate that for some genotypes the response stabilises already in the second week post-mutagenesis and for all in the third week.

14. Collect data on the plantlet height, fresh weight, survival, root length, number of shoots. For optimal results all these measurements should be performed on the same day for all treatments and replications to reduce bias. In the case of counting survival rate, count the number of healthy plantlets.

15. This calculation is made by dividing the weight of the mutated material (numerator) by the weight of the control material measured at the same time (denominator) multiplied by 100. For example, the weight of material treated is 1.46 g. The control material is 1.6 g. The percentage is then 1.46/1.60 x 100 = 91.25%.

16. At low doses fresh weight of irradiated material may exceed control values.

17. They may show different percent reductions at the same dose, but an overall trend can be derived by combining the data.

18. Keep in mind when planning later field trials that emergence and survival obtained under *in vitro* testing conditions may differ considerably from that under field conditions due to environmental stresses.

19. Jain (2010) published recommended doses for various banana cultivars: 10–20 Gy of gamma irradiation for diploid clones 'Calcutta 4' (AA genome) and 'Tani' (BB genome); 30–40 Gy of gamma irradiation for the triploids 'Three Hand Planty' (AAB genome), 'Grande Naine' (AAA genome), 'Williams' (AAA genome) and 'Kamaramasenge' (AAB genome); 40–50 Gy of gamma

irradiation for the triploid 'Cachaco' (ABB genome) (Jhurree-Dussoruth et al. 2014; Jain 2010).

20. Estimate the time needed to conduct the entire experiment with various tissue types. In case of time constrains and the high number of samples, split the plant material into batches for separate days, treating one tissue type per day. If this is being prepared a day before, place *in vitro* cuttings in a Petri plate with enough water. Induced mutations are random events, implying that even adherence to published irradiation conditions might not result in the same mutation events. A way of mitigating this uncertainty is to rely on statistical probability and to work with large population sizes. The optimal population size depends on the spectrum and density of induced mutations, and to a lesser extent on the application of the mutant population (forward or reverse genetics). A population size of 1000 M_1V_1 is easily accommodated in medium size laboratories and for polyploid bananas this may be considered suitable in reverse-genetics approaches to recover mutations in most target genes (Jankowicz-Cieslak et al. 2012). For forward-genetic screens a larger population size may be required because the chance of uncovering useful traits in polyploids in the M_1 generation is reduced due to heterozygous state of induced mutation. A guide is to target the production of a forward-genetic population of a minimum of 5000–10,000 individuals.

21. The first population after mutagenic treatment is referred to as M_1V_1 whereby V_1 signifies the first vegetative generation after mutagenic treatment. Increasing numbers following V represent successive vegetative generations and increasing numbers after M indicate meiotic propagation. This allows tracking of generations in both facultative and obligate vegetatively propagated species.

22. Ideally a line would represent a pedigree of clonally related material that begins with the first non-chimeric individual such that all subsequent material produced from this progenitor inherits the same mutations. Because the exact state where mutant *in vitro* plantlets are no longer chimeric is not easily determined, defining a line this way is not always possible. We therefore suggest defining a line simply as the pedigree of all material resulting from a single progenitor mutated plantlet. Materials in a line therefore may inherit the same mutations as observed in Jankowicz-Cieslak et al. (2012) but may also inherit different mutations if chimerism exists in meristematic cells at the time the progenitor is sub-cultured to produce daughter plantlets.

23. We recommend three sub-culturing cycles to ensure the dissolution of chimeric materials. If too few cycles are performed, the resulting plants may still be chimeric and mutations may not be stably passed from one to another vegetative generation. It is recommended that preliminary evaluation starts at M_1V_3, since this generation should have solid mutants whose uniformity must be assessed. Non-uniform mutant clones must undergo a further round of propagation to reach uniformity. In this generation the evaluation of desired traits, such as TR4

screening could be undertaken or delayed until more advanced generations. From M_1V_4 to M_1V_9 uniform clones may be propagated and planted in experimental trials to test their performance for desired traits such as biotic/abiotic stress. As early as in M_1V_4 generation, replicated trials of selected mutants may be conducted using parental or local varieties as checks. The M_1V_5 and M_1V_6 generations can be used in multi-locational trials and tested for performance in a range of environments and agronomic traits. Final assessment can be made in M_1V_9 to M_1V_{10} generations depending on plant species, the desired mutant clone or clones will be released as a new improved mutant variety (Spencer-Lopes et al. 2018).

24. It is possible to slow down the growth of plants at a temperature of 15–20 ° C and a reduced day cycle of 8–12 h light. Test lower growing temperature and light conditions if they are suitable before applying them to a new variety.

25. Shipment of plantlets: rooted plants in small, plastic culture tubes are very suitable for shipment. The medium containing gelling agent for single shoot development, is usually used for shipment.

26. Banana requires specific growth conditions during germination and growth. For hardening use a day cycle of 12 h at 24 °C and avoid direct sunlight. High ambient humidity is very important for the initial period of cultivation. Maintain relative humidity at more than 90% during the first week of weaning and thereafter within the range of 50% (minimum daytime humidity) to 90% (maximum night-time humidity) at all times, to reduce the rate of leaf drying (Vuylsteke and Talengera 1998). If no humidity facility is available, cover pots with plastic bags. Keep plantlets covered for 2-4 weeks for acclimatization.

27. Plantlets are kept in partial shade and can be acclimatized in the nursery for 2–3 months. During this period, off-types can be detected.

28. Initial morphometric analysis and can be evaluation for new phenotypes can be completed in the greenhouse at the whole plant level. Figure 1.3 illustrates the strategy for *in vitro* mutagenesis (from mutagenized shoot meristems), handling of the mutated population and mutation recovery in a vegetatively propagated plant. The isolated putative mutants in the M_1V_3 generation can be evaluated for stability and multiplied to test their agronomic performance. Tissue may be collected from M_1V_3 individuals, DNA extracted and screened (genotypically) for any induced mutations. The inheritance of isolated mutations, both genetic and phenotypic is evaluated and confirmed in the M_1V_6 and subsequent generations.

Acknowledgments Authors wish to thank Mr. Danilo Moreno for assistance in performing radiosensitivity experiments. Funding for this work was provided by the Food and Agriculture Organization of the United Nations and the International Atomic Energy Agency through their Joint FAO/IAEA Centre of Nuclear Techniques in Food and Agriculture. This work is part of IAEA Coordinated Research Project D22005.

References

Gosal SS, Wani SH, Kang MS (2010) Biotechnology and crop improvement. J Crop Improv 24: 153–217. ISSN: 1542-7528 print/1542-7535 online. https://doi.org/10.1080/15427520903584555

IAEA (2014) Radiation protection and safety of radiation sources: international basic safety standards, IAEA Safety Standards Series No. GSR Part 3, 2014. IAEA, Vienna, Austria. ISBN 978-92-0-135310-8

Jain SM (2010) In vitro mutagenesis in banana (Musa spp.) improvement. Acta Hortic 879:605–614. https://doi.org/10.17660/ActaHortic.2010.879.67

Jankowicz-Cieslak J. and Till J.B., 2016. Chemical mutagenesis of seed and vegetatively propagated plants using EMS. Current protocols in plant biology 1(4):617–635 (Wiley). DOI: https://doi.org/10.1002/cppb.20040

Jankowicz-Cieslak J, Huynh OA, Brozynska M, Nakitandwe J, Till BJ (2012) Induction, rapid fixation and retention of mutations in vegetatively propagated banana. Plant Biotechnol J 10: 1056–1066

Jhurree-Dussoruth B, Jankowicz-Cieslak J, Till BJ, Kallydin H, Burthia D (2014) Mutation induction and discovery in a Mauritian Dessert-type banana. Plant Gene Discovery & "Omics" Technologies, 16–19 February 2014, Vienna, Austria

Spencer-Lopes MM, Forster BP, Jankuloski L (eds) (2018) FAO/IAEA manual on mutation breeding – third edition. Food and Agriculture Organization of the United Nations, Rome, pp 251–266. ISBN 978-92-5-130526-3

Vuylsteke DR, Talengera D (1998) Post-flask management of micropropagated bananas and plantains. IITA, Ibadan, p 15

Chapter 2
Gamma Irradiation of Embryogenic Cell Suspension Cultures from Cavendish Banana (*Musa* spp. AAA Group) and *In Vitro* Selection for Resistance to Fusarium Wilt

Chunhua Hu, Yuanli Wu, and Ganjun Yi

Abstract In this chapter, the establishment of embryogenic cell suspension (ECS) cultures using immature male flowers of triploid banana (*Musa* AAA Cavendish subgroup cv. 'Brazil'), followed by somatic embryogenesis and plantlet regeneration is described. Mutation induction is achieved by exposing the ECS to gamma irradiation with the dose of 80 Gy. The mutagenized cell population is transferred to solid long-term suspension culture medium for 96 h to recover from mutagen treatment shock, followed by somatic embryo induction and development medium containing 20% crude culture filtrates from *Fusarium oxysporum* f. sp. *cubense* (Foc). After 90 days, the somatic embryos that survive are transferred to the germination medium containing 25% crude culture filtrates. The surviving mature somatic embryos are transferred to rooting medium after the fourth subculture on the germination medium containing 50% crude culture filtrates. Before transplanting in a Foc infected field, the *in vitro* plantlets are acclimatized and screened for resistance to Foc using a pot-based greenhouse bioassay.

Keywords Cavendish banana · Embryogenic callus · Gamma irradiation · Germplasm resistant to Fusarium wilt

C. Hu · Y. Wu · G. Yi (✉)
Institute of Fruit Tree Research, Guangdong Academy of Agricultural Sciences; Key Laboratory of South Subtropical Fruit Biology and Genetic Resource Utilization; Ministry of Agriculture and Rural Affairs, Guangdong Province Key Laboratory of Tropical and Subtropical Fruit Tree Research, Guangzhou, China
e-mail: yiganjun@vip.163.com

J. Jankowicz-Cieslak, I. L. Ingelbrecht (eds.), *Efficient Screening Techniques to Identify Mutants with TR4 Resistance in Banana*,
https://doi.org/10.1007/978-3-662-64915-2_2

1 Introduction

Bananas and plantains (*Musa* spp.) are not only the most widely consumed fruits in the world, but also a staple food for over 500 million people. In recent years, the world faced a sharp decline in banana production, due to extreme weather patterns and the outbreak of pests and diseases, especially Fusarium wilt. Today's global banana production is seriously threatened by a new strain of *Fusarium oxysporum* f. sp. *cubense* Tropical Race 4 (Foc TR4). There is an urgent need to introduce resistance against Foc TR4 into commercial banana cultivars. The most common cultivars for commercial production belong to *Musa* AAA Cavendish subgroup, which are sterile and seedless. Attempts to develop new banana genotypes resistant to Fusarium wilt using traditional cross breeding techniques face significant hurdles. Thus, induced mutagenesis or genetic engineering offers practical alternatives to create new varieties or novel germplasm and has become a dominant approach for breeding disease-resistant banana (Dita et al. 2018).

Somatic embryos originate from single cells, and the chimera frequency is very low. To accelerate genetic improvement of banana, it is important to establish embryogenic cell suspension (ECS) cultures, followed by plantlet regeneration through somatic embryogenesis. So far, different source explants have been used to establish ECS in banana, including basal leaf sheaths and corm sections (Novák et al. 1989), highly proliferating meristems (Dhed'a et al. 1991; Strosse et al. 2006), zygotic embryos (Marroquin et al. 1993), immature male (Escalant et al. 1994; Côte et al. 1996; Navarro et al. 1997; Becker et al. 2000; Pérez-Hernández and Rosell-García 2008; Kulkarni and Bapat 2012; Elayabalan et al. 2013; Namanya et al. 2014; Morais-Lino et al. 2016) and female flowers (Grapin et al. 2000). Among the various explants used, immature male flowers appear to be the most widely applicable starting material for the establishment of regenerable ECS.

Plant mutation breeding is an effective method for creating novel germplasm. Briefly, seeds, pollen, the whole plant, vegetative organs, or callus are subjected to irradiation, followed by the process of selection and identification of a new variety (Spencer-Lopes et al. 2018). Physical mutagenesis is also applied to improve horticultural crops, it is reported that novel germplasm with agriculturally valuable traits has been developed in apple, peach, pear, and citrus (Source: FAO/IAEA Mutant Varieties Database). Gamma irradiation has been widely used as a physical mutagen for breeding of many crops including banana (Novák et al. 1990; Mak et al. 1996; Guo et al. 2003).

In this chapter, we present a detailed protocol for the establishment of a cell suspension culture and plantlet regeneration *via* somatic embryogenesis of Cavendish banana (*Musa* spp. AAA group). The ECS cultures are then subjected to 80 Gy gamma irradiation, followed by *in vitro* selection for resistance to Fusarium wilt.

2 Materials

2.1 In Vitro *Media for Induction of Embryogenic Callus (MI), Long Term Suspension Culture (ML), Somatic Embryo Induction and Development (MSD), Somatic Embryo Germination (MG), and Rooting of Somatic Embryos (MR)*

1. MS (Murashige and Skoog 1962) basal medium.
2. MS basal medium without vitamins.
3. MS vitamins.
4. SH (Schenk and Hildebrandt 1972) basal medium without vitamins.
5. Morel and Wetmore (1951) vitamins.
6. Biotin.
7. Indole-3-acetic acid (IAA).
8. Naphthalene acetic acid (NAA).
9. 2, 4-Dichlorophenoxyacetic acid (2, 4-D).
10. 6-Benzyl aminopurine (6-BA).
11. Kinetin (KT).
12. Malt extract.
13. Glutamine.
14. Proline.
15. Sucrose.
16. Gelrite.
17. Agar.
18. Distilled water.
19. Analytical balance.
20. pH meter.
21. Autoclave.

2.2 *Materials for the Induction of Embryogenic Callus, Establishment of ECS Cultures, Induction and Maturation of Somatic Embryos*

1. Sterile culture media (MI, ML, MSD, MG, MR).
2. Male inflorescence of Cavendish banana (*Musa* spp. AAA group) (*see* Note 1).
3. 75% ethanol.
4. Sterile distilled water.
5. Stereo microscope.
6. Erlenmeyer flask (100 ml).
7. Petri dish (9.0 cm).
8. Sieve (154 μm and 900 μm).

9. Dissecting instruments (scalpels handle and blades, forceps).
10. Laminar airflow cabinet.
11. Rotary shaker.
12. Growth chamber with environmental control.

2.3 *Mutation Induction of ECS via Gamma-Irradiation*

1. Cobalt-60 source (0-100 Gy) (*see* Note 2).
2. ECS cultures.
3. Sterile ML agar medium.
4. Sterile filter paper.
5. Sterile petri dishes (9.0 cm).

2.4   **In Vitro** *Selection for Resistance to Fusarium Wilt*

1. Seven-days-old PDA (potato dextrose agar) plate culture of Foc TR4 strain II5 (NRRL 54006, VCG 01213).
2. Haemocytometer.
3. Czapek's Broth (ready to use).
4. Erlenmeyer flasks (250 ml).
5. Rotary shaker.
6. Cheesecloth.
7. Centrifuge.
8. Autoclave.
9. Microporous filters (0.25 μm).
10. Sterile MSD, MG and MR medium.
11. Substrates for *ex vitro* acclimatization of banana plantlets.
12. Greenhouse with environmental control.

3   Methods

3.1 *Preparation of MI, ML, MSD, MG, and MR Medium*

1. MI medium consists of MS basal medium, 1 mg/l biotin, 1 mg/l IAA, 1 mg/l NAA, 1 mg/l 2, 4-D, 100 mg/l glutamine, 100 mg/l malt extract, 30 g/l sucrose, and 7 g/l agar or 2 g/l gelrite.
2. ML medium consists of MS basal medium, 1 mg/l biotin, 1 mg/l 2,4-D, 100 mg/l glutamine, 100 mg/l malt extract, and 45 g/l sucrose.

3. MSD medium consists of SH basal medium without vitamins, MS vitamin, 1 mg/l
   biotin, 100 mg/l glutamine, 230 mg/l proline, 100 mg/l malt extract, 0.2 mg/l
   NAA, 0.1 mg/l KT, 45 g/l sucrose, 2 g/l gelrite.
4. MG medium consists of MS basal medium without vitamins, Morel and Wetmore
   vitamins, 1 mg/l 6-BA, 0.2 mg/l IAA, 30 g/l sucrose, 2 g/l gelrite.
5. MR medium is MG medium without any plant growth regulators.
6. Adjust pH value of MI and ML medium to 5.3, adjust pH value of other media
   to 5.8.
7. Sterilize all media for 15 min at 120 °C.
8. Allow media to cool prior to use.
9. Store for up to a week in a cold room.

3.2 Isolation of Immature Male Flowers

1. Take a female flower bud that has just completed the fruit set process and use the
   10-12 cm long portion at the top end of the bud, that is, the male inflorescence.
2. Keep on removing the outer bract and the male flower under it until the top 1.5 cm
   long portion of the inflorescence remains.
3. Surface sterilize the inflorescence by immersing in 75% ethanol for 1-2 min,
   followed by a rinse with sterile distilled water.
4. Under aseptic conditions, remove the inner bract and isolate the immature male
   flower under it with the aid of a stereo microscope. The one adjacent to the floral
   apex is rank 1 flower.
5. Rank 1 to rank 15 flowers are used as explant and are placed on MI medium.
6. The cultures are maintained at 28 ± 1 °C in darkness for nearly 5 months.

3.3 Induction of Embryogenic Callus and Development of ECS

1. For initiation of the suspension culture, select loose and fragile light-yellow callus
   induced on MI medium (see Fig. 2.1a).
2. Weigh ≈ 2 g of embryogenic callus and add to a 100 ml Erlenmeyer flask
   containing 30 ml ML medium. The cultures are incubated on a rotary shaker
   (110 rpm/min) at 28 ± 1 °C in darkness.
3. In the first month of suspension culture, replace ML medium once a week, and
   sieve cultures through a 900 μm mesh to remove the non-dispersible large culture
   particles, including callus clusters, dead tissues, and sometimes pre-embryos.
4. After one month, replace ML medium every 2 weeks. When cultures become
   dispersed, use a sieve of 154 μm aperture to remove the larger cell clusters.
5. Generally, it takes 3 months to obtain ECS cultures in Cavendish banana (Musa
   spp. AAA group), which are relatively more dispersed and homogeneous (see
   Fig. 2.1b).

Fig. 2.1 Steps for cell suspension culture and plantlet regeneration *via* somatic embryogenesis of Cavendish banana (*Musa* spp. AAA group cv. 'Brazil'). (**a**) Embryogenic callus induced on MI medium; (**b**) Well established embryogenic cell suspension; (**c**) Induction of somatic embryo on MSD medium; (**d**) Maturation of somatic embryo; (**e**) Germination of somatic embryo; (**f**) Regenerated plantlet

3.4 Induction and Maturation of Somatic Embryos

1. Ten days after subculture, cell clumps are taken from ECS cultures and sieved through a 154 μm mesh.
2. Transfer the cell population onto MSD medium (*see* Fig. 2.1c).
3. The cultures are maintained at 28 ± 1 °C in darkness.
4. Generally, it takes 90 days to promote maturation of somatic embryos (*see* Fig. 2.1d).

3.5 Plantlet Regeneration

1. Mature somatic embryos which have been cultured on MSD medium for 90 days are transferred to MG medium in Petri dishes.
2. The cultures are maintained at 28 ± 1 °C in darkness for 30 days.
3. After germination (*see* Fig. 2.1e), the somatic embryos with light green leaf sheaths are then transferred to MR medium for the development of complete plantlets (*see* Fig. 2.1f).
4. Cultures are maintained at 28 ± 1 °C under a 16 h/8 h photoperiod with 30 μmol m^{-2} s^{-1} from cool white fluorescent lamps.

3.6 Gamma Irradiation of ECS and In Vitro Selection for Resistance to Fusarium Wilt

3.6.1 Determination of Irradiation Dose

1. Prior to irradiation, the initial density of ECS is adjusted to 1.5% of the packed cell volume (PCV) in 30 ml of medium.
2. One week after subculture, ECS cultures are subjected to γ-ray irradiation treatment. Erlenmeyer flasks containing ECS cultures receive 40, 60, 80, and 100 Gy of irradiation, respectively, at a dose rate of 2 Gy/min. Control ECS cultures are prepared in the same manner but don't receive irradiation.
3. Weigh 0.1 g cell clumps and transfer to Erlenmeyer flask containing 40 ml MSD medium. Each treatment includes 4 replicates. The cultures are incubated in darkness for 90 days.
4. Weigh somatic embryo obtained, then transfer to MG medium. The cultures are incubated in darkness for 30 days.
5. Transfer the germinated somatic embryo to MR medium and calculate the number of regenerated plantlets.
6. Calculate LD_{50} according to the weight of somatic embryo and the number of regenerated plantlets (see Note 3).

3.6.2 Preparation of Crude Culture Filtrates from Foc TR4

1. Collect conidia from 7-days-old PDA plate culture of Foc TR4 strain by rinsing with sterile distilled water. Conidia concentration was determined with a haemocytometer and adjusted with sterile distilled water to 3×10^5—4×10^5 conidia /ml.
2. Take 1 ml of the conidia suspension and transfer to a 250 ml Erlenmeyer flask containing 100 ml Czapek's liquid medium. The culture is incubated on a rotary shaker (120 rpm/min) at $25 \pm 2\ °C$ for 12 days.
3. Then the culture is filtered through four layers of cheesecloth to remove mycelium, followed by filtrate being centrifuged at 5000 rpm/min for 15 min.
4. The supernatant broth is autoclaved at 121 °C for 20 min to eliminate the effect of enzymes, and then filtered under pressure with a bacteria filter with a 0.25 μm microfiltration membrane. The filtrate is the sterile crude culture filtrates from Foc.

3.6.3 In Vitro Selection for Resistance to Fusarium Wilt

1. Put a sterile filter paper on the surface of the ML agar medium, then transfer mutagen treated cells to the medium. The cultures are maintained for 96 h, which allows the cells to recover from mutagen treatment shock.

2. The mutagenized embryogenic cell population is transferred to MSD medium containing 20% crude culture filtrates from Foc.
3. After 90 days, the somatic embryos that survive are transferred to MG medium containing 25% crude culture filtrates.
4. After nearly 30 days, the surviving mature somatic embryos are then sub-cultured at least four times on the same medium containing 50% crude culture filtrates.
5. The regenerated shoots obtained are transferred to MR medium.
6. Before planting in a Foc infected field, the plantlets are acclimatized and screened for resistance to Foc using a pot system in greenhouse (see Note 4).

4 Notes

1. Although somatic embryogenesis in banana has been studied for more than 30 years (Cronauer-Mitra and Krikorian 1988), it is still far from being considered a routine technology. In other words, it is difficult to develop an efficient and repeatable protocol for all *Musa* genotypes. Given the fact that cultivars in Cavendish subgroup are the most common cultivars for commercial production around the world, this protocol describes culture establishment for a commercial cultivar 'Brazil' (*Musa* AAA Cavendish subgroup), which can also be applied to other Cavendish cultivars (Xu et al. 2003).
2. The mutagenesis was conducted in Guangdong Irradiation Center, which was only about an hour's ride from the institute. If the irradiation source is far away from the tissue culture lab, the risk of contamination during long-distance transportation require additional precautions, such as the use of dedicated and sterilized container for Erlenmeyer flasks and a temperature-controlled delivery vehicle.
3. It is observed that the growth and regeneration ability of ECS decreases gradually with the increase of irradiation dosage. After ECS cultures receive 80 Gy of irradiation, the weight of somatic embryo and the number of regenerated plantlets is 48.76% and 46.12% of the control, respectively. It is strongly recommended to determine the LD_{50} of the ECS cultures before ECS bulk irradiation.
4. The pot-based greenhouse screening protocol for *Musa* genotypes against Foc was performed as described by Zuo et al. (2018).

References

Becker DK, Dugdale B, Smith MK, Harding RM, Dale JL (2000) Genetic transformation of Cavendish banana (*Musa* spp. AAA group) cv 'Grand Nain' via microprojectile bombardment. Plant Cell Rep 19(3):229–234

Côte FX, Domergue R, Monmarson S, Schwendiman J, Teisson C, Escalant JV (1996) Embryogenic cell suspensions from the male flower of *Musa* AAA cv. 'Grand Nain'. Physiol Plant 97: 285–290

Cronauer-Mitra SS, Krikorian AD (1988) Plant regeneration via somatic embryogenesis in the seeded diploid banana *Musa ornata* Roxb. Plant Cell Rep 7:23–25

Dhed'a D, Dumortier F, Panis B, Vuylsteke D, Langhe ED (1991) Plant regeneration in cell suspension cultures of the cooking banana cv. 'Bluggoe' (*Musa* spp. ABB group). Fruits 46(2):125–135

Dita M, Barquero M, Heck D, Mizubuti ESG, Staver CP (2018) Fusarium wilt of banana: current knowledge on epidemiology and research needs toward sustainable disease management. Front Plant Sci 9:1468

Elayabalan S, Kalaiponmani K, Pillay M, Chandrasekar A, Selvarajan R, Kumar KK, Balasubramanian P (2013) Efficient regeneration of the endangered banana cultivar 'Virupakshi' (AAB) via embryogenic cell suspension from immature male flowers. Afr J Biotechnol 12:563–569

Escalant JV, Teisson C, Cote F (1994) Amplified somatic embryogenesis from male flowers of triploid banana and plantain cultivars (*Musa* spp.). In Vitro Cellular Dev Biol Plant 30(4): 181–186

Grapin A, Ortíz JL, Lescot T, Ferrière N, Côte FX (2000) Recovery and regeneration of embryogenic cultures from female flowers of false horn plantain. Plant Cell Tissue Organ Culture 61(3): 237–244

Guo JH, Cai EX, Lin QT, Chen LP , Huang XD, Shen MS (2003) Study on mutation breeding of banana buds *in vitro* IV: bio-chemical analysis to 'Zhangjiao No. 8' strain. Subtropical Plant Sci 32(1):11–13. (in Chinese)

Kulkarni VM, Bapat VA (2012) Somatic embryogenesis and plant regeneration from cell suspension cultures of Rajeli (AAB), and endangered banana cultivar. J Plant Biochem Biotechnol 22: 132–137

Mak C, Ho YW, Tan YP, Rusli I (1996) 'Novaria' - a new banana mutant induced by gamma irradiation. Infomusa 5:35–36

Marroquin CG, Paduscheck C, Escalant JV, Teisson C (1993) Somatic embryogenesis and plant regeneration through cell suspensions in *Musa acuminata*. In Vitro Cellular Dev Biol Plant 29: 43–46

Morais-Lino LS, Santos-Serejo JA, Amorim EP, de Santana JRF, Pasqual M, de Oliveira e Silva S (2016) Somatic embryogenesis, cell suspension, and genetic stability of banana cultivars. In Vitro Cellular Dev Biol Plant 52:99–106

Morel G, Wetmore RH (1951) Tissue culture of monocotyledons. Am J Bot 38:138–140

Murashige T, Skoog F (1962) A revised medium for rapid growth and bioassays with tobacco tissue cultures. Physiol Plant 15:473–497

Namanya P, Mutumba G, Magambo SM, Tushemereirwe W (2014) Developing a cell suspension system for *Musa*-AAA-EA cv. 'Nakyetengu': a critical step for genetic improvement of Matooke East African Highland bananas. In Vitro Cellular Dev Biol Plant 50:442–450

Navarro C, Escobedo RM, Mayo A (1997) *In vitro* plant regeneration from embryogenic cultures of a diploid and a triploid, Cavendish banana. Plant Cell Tissue Organ Culture 51(1):17–25

Novák FJ, Afza R, van Duren M, Perea-Dallos M, Conger BV, Tang XL (1989) Somatic embryogenesis and plant regeneration in suspension cultures of dessert (AA and AAA) and cooking (ABB) bananas (*Musa* spp.). Bio/Technology 7:154–159

Novák FJ, Afza R, van Duren M, Omar MS (1990) Mutation induction by gamma irradiation of *in vitro* cultured shoot-tips of banana and plantain (*Musa* cvs.). Trop Agric 67:21–28

Pérez-Hernández JB, Rosell-García P (2008) Inflorescence proliferation for somatic embryogenesis induction and suspension-derived plant regeneration from banana (*Musa* AAA, cv. 'Dwarf Cavendish') male flowers. Plant Cell Rep 27:965–971

Schenk RU, Hildebrandt AC (1972) Medium and techniques for induction and growth of monocotyledonous and dicotyledonous plant cell cultures. Can J Bot 50:199–204

Spencer-Lopes MM, Forster BP, Jankuloski L (eds) (2018) Manual on mutation breeding, 3rd edn. Food and Agriculture Organization of the United Nations/International Atomic Energy Agency, Vienna

Strosse H, Schoofs H, Panis B, Andre E, Reyniers K, Swennen R (2006) Development of embryogenic cell suspensions from shoot meristematic tissue in bananas and plantains (*Musa* spp.). Plant Sci 170(1):104–112

Xu CX, Li HP, Xiao HG, Fan HZ (2003) Establishment of embryogenic cell suspensions from meristematic globules of *Musa* spp. Acta Horticulturae Sinica 30:580–582 (in Chinese)

Zuo CW, Deng GM, Li B, Huo HQ, Li CY, Hu CH, Kuang RB, Yang QS, Dong T, Sheng O, Yi GJ (2018) Germplasm screening of *Musa* spp. for resistance to *Fusarium oxysporum* f. sp. *cubense* tropical race 4 (*Foc* TR4). Eur J Plant Pathol 151:723–734

Part II
Fusarium TR4 Screening Technologies

Chapter 3
Pre-Screening of Banana Genotypes for Fusarium Wilt Resistance by Using an *In Vitro* Bioassay

Yuanli Wu and Ganjun Yi

Abstract In the process of breeding and selection of banana for resistance to Fusarium wilt, it is important to conduct an efficient resistance screening test by artificial inoculation with *Fusarium oxysporum* f. sp. *cubense* (Foc) Tropical Race 4. So far, there are two types of early bioassays for screening *Musa* genotypes against Foc: a greenhouse and an *in vitro* bioassay. The most commonly used greenhouse bioassay is a pot-based system followed by a hydroponic system. Here we describe an *in vitro* bioassay characterized by *in vitro* inoculation of rooted banana plantlets grown on medium consisting of half-strength MS macronutrients and MS micronutrients. The disease response and evaluation results obtained through this *in vitro* bioassay correlates with that from a greenhouse screen and/or field evaluation. Given the importance of *in vitro* cell and tissue culture techniques for banana (mutation) breeding, promising resistant clones could be screened directly. This *in vitro* bioassay is a totally contained system compared with greenhouse methods and does not require an acclimatization step, thereby improving banana breeding efficiency. The *in vitro* pre-screening protocol and bioassay for Fusarium wilt resistance presented here is fast, space-effective, and accurate.

Keywords Bioassay · *Fusarium oxysporum* f. sp. *cubense* · *Musa* · Resistance

1 Introduction

Fusarium wilt or Panama disease caused by the pathogenic fungus *Fusarium oxysporum* f. sp. *cubense* (Foc) is one of the most destructive diseases of banana and is found in all areas where banana is grown (Ploetz 2015). The soil-borne

Y. Wu · G. Yi (✉)
Institute of Fruit Tree Research, Guangdong Academy of Agricultural Sciences; Key Laboratory of South Subtropical Fruit Biology and Genetic Resource Utilization; Ministry of Agriculture and Rural Affairs, Guangdong Province Key Laboratory of Tropical and Subtropical Fruit Tree Research, Guangzhou, China
e-mail: yiganjun@vip.163.com

© The Author(s) 2022
J. Jankowicz-Cieslak, I. L. Ingelbrecht (eds.), *Efficient Screening Techniques to Identify Mutants with TR4 Resistance in Banana*,
https://doi.org/10.1007/978-3-662-64915-2_3

pathogen infects the roots of banana plants and colonizes the xylem vessels, which leads to typical external symptoms such as wilting and yellowing of the foliage, eventually leading to plant mortality. Currently, there is still a lack of economically viable measures for managing Fusarium wilt in an infected field (Dita et al. 2018). It is generally accepted that breeding and selection for disease resistance is the only effective and sustainable management option.

Promising resistant clones acquired through conventional and non-conventional breeding techniques should be screened for resistance to Fusarium wilt using artificial inoculation with Foc. So far, there are two types of early bioassays for screening *Musa* genotypes against Foc: a greenhouse and an *in vitro* bioassay (Wu et al. 2010; Ghag et al. 2012; Hu et al. 2013). The most commonly used greenhouse bioassay is a pot system (Dita et al. 2011; Ribeiro et al. 2011; Li et al. 2015; Reboucas et al. 2018; Zhang et al. 2018; Zuo et al. 2018) followed by a hydroponic system.

Although some progress has been achieved, attempts at developing new banana genotypes resistant to Fusarium wilt using conventional breeding techniques face significant hurdles, mainly because most cultivars of *Musa* AAA Cavendish subgroup are sterile and seedless (Ortiz 2013). Nowadays, non-conventional breeding approaches for banana improvement such as somaclonal variation and genetic transformation have received more attention. Somaclonal variation caused by long-term *in vitro* propagation is considered an important source of genetic variability, through which several tolerant clones have been acquired (Hwang and Ko 2004). In addition, a genetic transformation protocol has been well established in different banana genotypes, which can be used to create transgenic plants resistant to Foc Tropical Race 4 (Foc TR4) (Dale et al. 2017). Given the fact that non-conventional breeding techniques are based on banana cell and tissue culture, promising resistant clones could be screened directly using an *in vitro* bioassay. Based on the modification of previous work (Wu et al. 2010), we present here a pre-screening protocol for Fusarium wilt resistance by using an *in vitro* bioassay that is fast, space-effective, and accurate.

The *in vitro* bioassay is characterized by *in vitro* inoculation of rooted banana plantlets grown in a half-strength MS medium without a carbon source. Twenty-four days after inoculation with Foc TR4 at 10^6 conidia/ml, the disease severity was rated on a scale from 1 to 6 (Wu et al. 2010). Results of the disease score were then subjected to ordinal logit model analysis, which is also known as proportional odds model (McCullagh 1980). According to symptom rating probability distribution, the reaction of banana genotypes against Fusarium wilt was divided into five categories: highly resistant (HR), resistant (R), moderately resistant (MR), susceptible (S), and highly susceptible (HS). Compared with the greenhouse bioassay, this *in vitro* bioassay is a totally closed system. Since acclimatization of *in vitro* plantlets is not required, the application of the bioassay improves banana breeding efficiency.

2 Materials

2.1 Medium for Interaction System (MIS)

1. MS (Murashige and Skoog 1962) macronutrients and micronutrients (*see* Note 1).
2. Tissue culture grade water.
3. Gelling agent (*e.g.* agar).
4. Analytical balance.
5. Weighing trays.
6. Spatula.
7. Beakers (500 ml and 1000 ml).
8. Magnetic stir bar.
9. Hot plate.
10. pH meter.
11. NaOH (1 N).
12. HCl (1 N).
13. Erlenmeyer flasks (150 ml).
14. Ventilated sealing film for Erlenmeyer flask (12 × 12 cm).
15. Rubber bands.

2.2 Plant Material Preparation

1. High-quality *in vitro* plantlets which have been cultured on rooting medium for 2 weeks (Wu et al. 2005).
2. Sterile MIS medium.
3. Forceps.
4. Scalpel handle.
5. Scalpel blades.

2.3 Inoculum Preparation

1. Five-days-old PDA (potato dextrose agar) plate culture of Foc TR4 strain GD-13 (VCG 01213/16, ACCC 37997), which was isolated from a diseased Cavendish banana plant in Guangdong Province, PR China (*see* Note 2).
2. Electron Microscopy Sciences (EMS) Rapid-Core (6.0 mm).
3. Potato Dextrose Water (ready-to-use).
4. Erlenmeyer flask (500 ml).

5. Orbital shaker.
6. Cheesecloth.
7. Petri dishes.
8. Short-stem funnel (60 mm).
9. Bottles (100 ml).
10. Beakers (500 ml).
11. Forceps.
12. Haemocytometer.

2.4 *In Vitro* Inoculation

1. Filter paper.
2. Petri dish.
3. Sterile distilled water.
4. Sterile centrifuge tube (50 ml).
5. Beakers (100 ml).
6. Forceps.
7. Disposable pipettes.

3 Methods (*See* Note 3)

3.1 *Preparation of MIS Medium*

1. Prepare stock solutions of MS macronutrients (20×) and micronutrients (200×).
2. For 1 liter of MIS medium, weigh 6 g of agar in 400 ml of tissue culture grade water.
3. Heat while stirring until the solution is homogenous and clear.
4. In a separate beaker containing 400 ml water mix 25 ml stock solution of MS macronutrients (20x) and 5 ml stock solution of MS micronutrients (200x), stir the mixture.
5. Mix the solution prepared in step 3 with the solution prepared in step 4.
6. Add water to a final volume of 1 liter, while stirring adjust the pH to 5.8 using NaOH and HCl.
7. Dispense 50 ml of the medium into Erlenmeyer flasks.
8. Close each Erlenmeyer flask over the top with a ventilated sealing film and a rubber band.
9. Sterilize the medium for 20 min at 120 °C.
10. Allow medium to cool prior to use.
11. Store for up to a week in a cold room.

3.2 Plant Material Preparation

1. Autoclave forceps and scalpel.
2. In a laminar flow bench, remove the blackened part at the base of the rooted plantlets (the height of the pseudostem is 3.5–4.0 cm), shorten the roots to approximately 0.5 cm in length.
3. Transfer plantlets into Erlenmeyer flasks containing MIS, one plantlet per flask.
4. Maintain cultures at $25 \pm 2\ °C$ under a 12 h photoperiod with $50\ \mu mol\ m^{-2}\ s^{-1}$ from cool white fluorescent lamps.
5. After 1–2 weeks of growth, select plantlets for *in vitro* inoculation: the height of the pseudostem should measure 4.5–5.0 cm, and the plantlet should possess more than two fully expanded leaves and at least three white roots.

3.3 Inoculum Preparation

1. Cut cheesecloth into round pieces (8 cm) and then place in a glass Petri dish.
2. Autoclave forceps, Erlenmeyer flask, Petri dish with cheesecloth pieces, short-stem funnel, bottle, and beaker.
3. In a laminar flow bench, dispense 200 ml of Potato Dextrose Water into Erlenmeyer flasks.
4. Punch out 5 mycelial plugs from the outer edge of the actively growing colony of Foc TR4 strain and transfer into an Erlenmeyer flask containing Potato Dextrose Water.
5. Place the Erlenmeyer flask on a rotary shaker (150 rpm) and incubate at 25 °C for 5 days.
6. In a laminar flow bench, insert funnel into bottle, then place four layers of cheesecloth into funnel.
7. Pour sporulation medium through four layers of cheesecloth to separate spores from mycelia.
8. After filtration, place cheesecloth pieces and funnel into beaker.
9. Determine conidia concentration in spore suspension using a haemocytometer.
10. Store spore suspension at 4 °C until *in vitro* inoculation.
11. Autoclave beaker with cheesecloth pieces and funnel.
12. Allow beaker to cool prior to disposing cheesecloth pieces.

3.4 In Vitro Inoculation (See Fig. 3.1 and Note 4)

1. Cut filter paper into small discs (5 mm) and then place in a glass Petri dish.
2. Autoclave forceps, Petri dish with filter paper discs, disposable pipettes, and beaker.

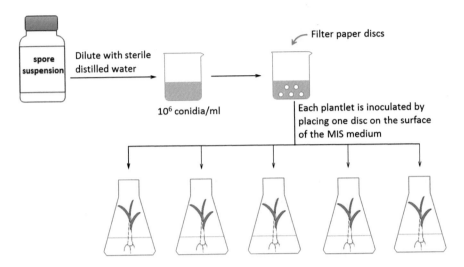

Fig. 3.1 *In vitro* inoculation protocol

3. In a laminar flow bench, adjust conidia concentration to 10^6 conidia/ml with sterile distilled water in a sterile centrifuge tube.
4. Pour the spore suspension (10^6 conidia/ml) into a beaker, then soak the filter paper discs in the spore suspension.
5. Inoculate each plantlet with Foc TR4 by placing one disc on the surface of the MIS medium.
6. Maintain cultures at $25 \pm 2\,°C$ under a 12 h photoperiod with 50 μmol m^{-2} s^{-1} from cool white fluorescent lamps for 24 days.
7. Autoclave beaker containing spore suspension for decontamination.

3.5 Disease Severity Rating and Statistical Analysis

1. Rate disease severity of the *in vitro* inoculated plantlets on a scale of 1 to 6: 1, no discoloration of the pseudostem; 2, $\leq$ 1/2 the height of the pseudostem discolored; 3, >1/2 the height of the pseudostem discolored and (or) leaf stalk discolored; 4, $\leq$50% of the leaves wilted or yellowed; 5, >50% of the leaves wilted or yellowed; and 6, the whole plantlet wilted (*see* Fig. 3.2).
2. After 24 days incubation assess disease severity by determining disease score.
3. Data collection: record the data for each banana plantlet and enter it into a spreadsheet (*e.g.* Microsoft Excel).
4. Model construction based on cumulative Logistic regression (*see* Note 5). Wherein the Logistic regression model accordingly contains five logit functions:

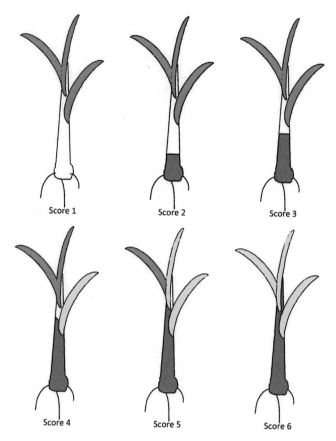

Fig. 3.2 A scale of 1–6 is used to measure disease severity of banana rooted plantlets 24 days after *in vitro* inoculation with *Fusarium oxysporum* f. sp. *cubense* tropical race 4 at 10^6 conidia/ml

$$\ln\left(\frac{p_1}{1-p_1}\right) = \beta_{01} - \sum_{k=1}^{k}\beta_k x_k$$

$$\ln\left(\frac{p_1+p_2}{1-p_1-p_2}\right) = \beta_{02} - \sum_{k=1}^{k}\beta_k x_k$$

$$\ln\left(\frac{p_1+p_2+p_3}{1-p_1-p_2-p_3}\right) = \beta_{03} - \sum_{k=1}^{k}\beta_k x_k$$

$$\ln\left(\frac{p_1+p_2+p_3+p_4}{1-p_1-p_2-p_3-p_4}\right) = \beta_{04} - \sum_{k=1}^{k}\beta_k x_k$$

$$\ln\left(\frac{p_1 + p_2 + p_3 + p_4 + p_5}{p_6}\right) = \beta_{05} - \sum_{k=1}^{k} \beta_k x_k$$

Wherein p_1, p_2, p_3, p_4, p_5 and p_6 are event probabilities, which respectively represent disease grades of 1–6, and the basal level for comparison is grade 6; x_k ($k = 1,2\dots$, K) represents banana cultivar; β_{0j} ($j = 1,2\dots$, 5) represents an intercept term of regression; and β_k ($k = 1,2\dots$, K) represents a regression coefficient; each logit function has the same coefficient term and different intercept terms, and the regression lines of each cumulative logit are parallel to each other.

The estimation method used for the Logistic regression model is a maximum likelihood method, according to the aforementioned Logistic model function designed for predicting the disease severity of the rooted plantlet of banana, the accumulated Logistic regression model obtained is described as follows:

$$y' = \alpha + \sum_{k=1}^{K} \beta_k x_k + \varepsilon$$

Wherein y' represents the disease incidence of the rooted plantlet of banana, α represents an intercept term; β_k ($k = 1,2\dots$, K) represents a regression coefficient; x_k ($k = 1,2\dots$, K) represents banana cultivar, and ε is an error term.

5. Calculation of cumulative probability: assigning respective values $y = 1$, $y = 2$,...., $y = 6$ to six disease grades, wherein a relationship among individual y values is $(y = 1) < (y = 2) < \dots < (y = 6)$, and there are 5 demarcation lines for demarcating adjacent categories:

if $y' \le \mu_1$, $y = 1$
if $\mu_1 < y' \le \mu_2$, $y = 2$
if $\mu_2 < y' \le \mu_3$, $y = 3$
if $\mu_3 < y' \le \mu_4$, $y = 4$
if $\mu_4 < y' \le \mu_5$, $y = 5$
if $\mu_5 < y'$, $y = 6$

μ_j is a demarcation point for demarcating categories, and $\mu_1 < \mu_2 < \mu_3 < \mu_4 < \mu_5$.

The formula for calculating a cumulative probability value is as follows:

$$P(y \le j|x) = P(y' \le \mu_j|x) = \frac{1}{1 + e^{-\left[\mu_j - \left(\alpha + \sum_{k=1}^{k} \beta_k x_k\right)\right]}}$$

Therefore, the probability value of the rooted plantlet of a banana cultivar in a certain disease grade can be obtained:

$$P\left(y=1\right)=P\left(y\leq1\right)$$
$$P\left(y=2\right)=P\left(y\leq2\right)-P\left(y\leq1\right)$$
$$P\left(y=3\right)=P\left(y\leq3\right)-P\left(y\leq2\right)$$
$$P\left(y=4\right)=P\left(y\leq4\right)-P\left(y\leq3\right)$$
$$P\left(y=5\right)=P\left(y\leq5\right)-P\left(y\leq4\right)$$
$$P\left(y=6\right)=1-P\left(y\leq5\right)$$

The sum of the probability values of each grade is 1, that is, $P\left(y=1\right)+P\left(y=2\right)+\ldots+P\left(y=6\right)=1$.

6. Open the spreadsheet at user interface of statistic software (*e.g.* IBM SPSS Statistics), then execute ordinal logit model analysis and obtain symptom rating probability distribution of each banana cultivar or clone. As an example, the result of statistical analysis is given in Fig. 3.3.

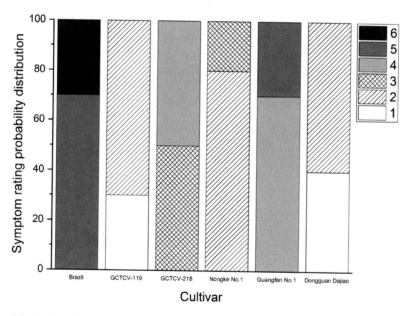

Fig. 3.3 *In vitro* inoculation of six banana cultivars with *Fusarium oxysporum* f. sp. *cubense* Tropical Race 4 (Foc TR4). Twenty-four days after *in vitro* inoculation of rooted plantlets (*n* = 15), disease symptoms were scored on an ordinal scale as illustrated: 1, no discoloration of the pseudostem; 2, ≤1/2 the height of the pseudostem discolored; 3, >1/2 the height of the pseudostem discolored and (or) leaf stalk discolored; 4, ≤50% of the leaves wilted or yellowed; 5, >50% of the leaves wilted or yellowed; and 6, the whole plantlet wilted. Control plantlets were treated in the same manner except that Foc TR4 was replaced with sterile distilled water. Data obtained from three independent experiments were subjected to ordinal logit model analysis, which is also known as proportional odds model

Table 3.1 Criteria for grading resistance against *Fusarium oxysporum* f. sp. *cubense*

Symptom rating probability distribution	Level of resistance to Fusarium wilt
$P(y=2) \geq 50\%$, $P(y=1) \leq 50\%$	HR
$P(y=2) \geq 50\%$, $P(y=3) \leq 50\%$	R
$P(y=3) \geq 50\%$, $P(y=2) <50\%$ or $P(y=4) \leq 50\%$	MR
$P(y=4) \geq 50\%$, $P(y=3) <50\%$ or $P(y=5) \leq 50\%$	S
$P(y=5) \geq 50\%$, $P(y=4) <50\%$ or $P(y=6) \leq 50\%$	HS

Table 3.2 Screening of banana cultivars for resistance to Fusarium wilt under field, greenhouse, and *in vitro* conditions, respectively

	Level of resistance to Fusarium wilt		
Cultivar	Field Evaluation	Greenhouse Bioassay	*In vitro* Bioassay
Brazil (*Musa* AAA Cavendish subgroup)	HS (Huang et al. 2005)	S (Zuo et al. 2018)	HS
GCTCV-119 (*Musa* AAA Cavendish subgroup)	HR (Huang et al. 2005)	R (Zuo et al. 2018)	HR
GCTCV-218 (*Musa* AAA Cavendish subgroup)	MR (Huang et al. 2005)	–	MR
Nongke No.1 (*Musa* AAA Cavendish subgroup)	R (Liu et al. 2007)	–	R
Guangfen No.1 (*Musa* ABB Pisang Awak)	–	S (Zuo et al. 2018)	S
Dongguan Dajiao (*Musa* ABB group)	HR (Huang et al. 2005)	–	HR

7. Evaluate the disease resistance level of banana cultivars or clones according to their symptom rating probability distributions (*see* Table 3.1). As an example, the evaluation result is given in Table 3.2.
8. Autoclave Erlenmeyer flasks with *in vitro* inoculated plantlets for decontamination (*see* Notes 6 and 7).

4 Notes

1. MIS medium is composed of half-strength MS macronutrients and MS micronutrients without a carbon source. Media commonly used for the growth and sporulation of Foc are carbohydrate-rich, and carbon has been shown to be the first limiting substrate of Foc growth in sterilized soil (Couteaudier and Alabouvette 1990). In this bioassay, the subtle balance between the growth of Foc and the carbon source is achieved by removing sucrose from the MS medium

and by employing a filter paper disc as carbon source. Since the rooted banana plantlets with leaves are able to photosynthesize, the use of MIS medium in this bioassay also guarantees the normal growth of the plantlets.

2. CAUTION: it is suggested to use a virulent strain of Foc TR4. The disease severity of *in vitro* inoculated plantlets is affected by several factors, such as the pathogenicity of the Foc TR4 strain, sporulation medium, components of MIS medium, and growing time of banana plantlet on MIS medium before *in vitro* inoculation. Among these, the pathogenicity of Foc TR4 strain is the most important factor (Wu et al. 2020). If variation in pathogenicity occurs, for example, due to repeated sub-culturing or long-term preservation in the lab, this bioassay is not useful to identify banana mutants which are resistant to Fusarium wilt.

3. CAUTION: not only plant material preparation but also inoculum preparation and inoculation of rooted banana plantlets are conducted under sterile conditions. In order to obtain reliable data of disease severity, it's necessary to acquire the ability of aseptic operation prior to the experiment.

4. Five plantlets of each cultivar are *in vitro* inoculated with Foc TR4, and the entire experiment is repeated three times ($n = 15$). Control plantlets are treated in the same manner except that filter paper discs are soaked in sterile distilled water.

5. For field evaluation and greenhouse bioassay, the resistance of banana cultivars tested is generally divided into five levels according to a disease index. Because the disease grades of *in vitro* bioassay are non-linear in disease severity, the numerical value obtained by substituting the disease index calculation formula cannot reflect the real disease severity. That is, when there is no quantitative limit for disease grade, the disease index calculation formula is not applicable. The logistic regression model belongs to a probabilistic nonlinear regression model (McCullagh 1980). Using Logistic regression to analyze the disease grade data greatly improves the accuracy of the *in vitro* bioassay.

6. The *in vitro* bioassay is a laboratory bioassay with effective containment conducted in a controlled environment. All experimental waste and materials must be autoclaved in order to prevent the spread of Fusarium wilt pathogen into the environment.

7. Since tissue-cultured banana plantlets get easily contaminated during plant material preparation and *in vitro* inoculation, it is necessary to re-isolate Foc TR4 by cutting a thin cross-section of the rhizome from a diseased plantlet followed by placing the disc directly onto a PDA plate. Then the cultures are identified by following the procedures described by Summerell et al. (2003).

References

Couteaudier Y, Alabouvette C (1990) Survival and inoculum potential of conidia and chlamydospores of *Fusarium oxysporum* f. sp. *lini* in soil. Can J Microbiol 36(8):551–556

Dale J, James A, Paul JY, Khanna H, Smith M, Peraza-Echeverria S, Garcia-Bastidas F, Kema G, Waterhouse P, Mengersen K, Harding R (2017) Transgenic Cavendish bananas with resistance to Fusarium wilt tropical race 4. Nat Commun 8:1496

Dita MA, Waalwijk C, Paiva LV, Souza Jr MT, Kema GHJ (2011) A greenhouse bioassay for the *Fusarium oxysporum* f. sp. *cubense* x 'Grand Naine' (*Musa*, AAA, Cavendish subgroup) interaction. Acta Hortic 897:377–380

Dita M, Barquero M, Heck D, Mizubuti ESG, Staver CP (2018) Fusarium wilt of banana: current knowledge on epidemiology and research needs toward sustainable disease management. Front Plant Sci 9:1468

Ghag SB, Shekhawat UKS, Ganapathi TR (2012) Petunia floral defensins with unique prodomains as novel candidates for development of Fusarium wilt resistance in transgenic banana plants. PLoS One 7:e39557

Hu CH, Wei YR, Huang YH, Yi GJ (2013) An efficient protocol for the production of *chit42* transgenic Furenzhi banana (*Musa* spp. AA group) resistant to *Fusarium oxysporum*. In Vitro Cellular Dev Biol Plant 49(5):584–592

Huang BZ, Xu LB, Yang H, Tang XL, Wei YR, Qiu JS, Li GQ (2005) Preliminary results of field evaluation of banana germplasm resistant to Fusarium wilt disease. Guangdong Agric Sci 32(6):9–10. (in Chinese)

Hwang SC, Ko WH (2004) Cavendish banana cultivars resistant to Fusarium wilt acquired through somaclonal variation in Taiwan. Plant Dis 88(6):580–588

Li WM, Dita M, Wu W, Hu GB, Xie JH, Ge XJ (2015) Resistance sources to *Fusarium oxysporum* f. sp. *cubense* tropical race 4 in banana wild relatives. Plant Pathol 64(5):1061–1067

Liu SQ, Liang ZH, Huang ZH, Huang YX (2007) Selection of 'Nongke No.1' (*Musa* AAA Cavendish subgroup) resistant to Fusarium wilt. Guangdong Agric Sci 34(1):30–32. (in Chinese)

McCullagh P (1980) Regression models for ordinal data (with discussion). J R Stat Soc Ser B 42(2): 109–142

Murashige T, Skoog F (1962) A revised medium for rapid growth and bioassays with tobacco tissue cultures. Physiol Plant 15:473–497

Ortiz R (2013) Conventional banana and plantain breeding. Acta Hortic 986:177–194

Ploetz RC (2015) Fusarium wilt of banana. Phytopathology 105(12):1512–1521

Reboucas TA, Haddad F, Ferreira CF, de Oliveira SAS, da Silva Ledo CA, Amorrim EP (2018) Identification of banana genotypes resistant to Fusarium wilt race 1 under field and greenhouse conditions. Sci Hortic 239:308–313

Ribeiro LR, Amorim EP, Cordeiro ZJM, de Oliveira e Silva S, Dita MA (2011) Discrimination of banana genotypes for Fusarium wilt resistance in the greenhouse. Acta Hortic 897:381–385

Summerell BA, Salleh B, Leslie JF (2003) A utilization approach to *Fusarium* identification. Plant Dis 87:117–128

Wu YL, Yi GJ, Yang H, Zhou BR, Zeng JW (2005) Basal medium with modified nitrogen source and other factors influence the rooting of banana. HortScience 40(2):428–430

Wu YL, Yi GJ, Peng XX (2010) Rapid screening of *Musa* species for resistance to Fusarium wilt in an *in vitro* bioassay. Eur J Plant Pathol 128(3):409–415

Wu YL, Huang BZ, Zhang ZS, Yang XY (2020) Modification of *in vitro* bioassay for screening *Musa* species against *Fusarium oxysporum* f. sp. *cubense*. Acta Horticulturae Sinica 47(8): 1577–1584. (in Chinese)

Zhang L, Yuan TL, Wang YZ, Zhang D, Bai TT, Xu ST, Wang YY, Tang WH, Zhang SJ (2018) Identification and evaluation of resistance to *Fusarium oxysporum* f. sp. *cubense* tropical race 4 in *Musa acuminata* Pahang. Euphytica 214(7):106

Zuo CW, Deng GM, Li B, Huo HQ, Li CY, Hu CH, Kuang RB, Yang QS, Dong T, Sheng O, Yi GJ (2018) Germplasm screening of *Musa* spp. for resistance to *Fusarium oxysporum* f. sp. *cubense* tropical race 4 (*Foc* TR4). Eur J Plant Pathol 151(3):723–734

Chapter 4

In Vitro Based Mass-Screening Technique for Early Selection of Banana Mutants Resistant to Fusarium Wilt

Behnam Naserian Khiabani

Abstract Banana and plantains are among the most valuable agricultural commodities in the world. Banana Fusarium wilt, caused by the soil-borne *Fusarium oxysporum* f. sp. *cubense* (Foc), is one of the most devastating diseases of banana globally. In the 1990s a new strain of *Fusarium oxysporum* called tropical race 4 (TR4) emerged in Southeast Asia that affected commercial Cavendish plantations. The development of resistant cultivars is an effective strategy for management of the disease. Field-based screening to identify Foc-resistant plants is time-consuming, expensive and is often challenged by variable environmental conditions. Here we present an early selection protocol enabling evaluation of the disease under *in vitro* conditions. This method provides a preliminary screening and allows evaluation of a large number of *in vitro* plantlets. Using this method, within a short time and in a small laboratory, breeders can evaluate thousands of banana plantlets, produced *via* irradiation. Subsequently, putative, disease-resistant mutant lines can be identified and evaluated in the field.

Keywords Banana · *Fusarium oxysporum* · *In vitro* bioassay · Mass selection

1 Introduction

Banana and plantains belong to the genus *Musa* and are important agricultural products in developing countries. More than 1000 varieties of bananas are produced and consumed locally. Cavendish banana (AAA) are the main commercial variety for export and international trade and account for around 47% of global production (FAO 2019a). Approximately 50 million tons of Cavendish bananas are being produced globally every year. In 2017 the global banana production reached

B. N. Khiabani (✉)
Plant Breeding Department, Nuclear Agriculture Research School, Nuclear Science and Technology Research Institute (NSTRI), Tehran, Iran
e-mail: bnaserian@aeoi.org.ir

© The Author(s) 2022
J. Jankowicz-Cieslak, I. L. Ingelbrecht (eds.), *Efficient Screening Techniques to Identify Mutants with TR4 Resistance in Banana*,
https://doi.org/10.1007/978-3-662-64915-2_4

114 million tons (FAO 2019b). Bananas are locally consumed as vital staple food or as a significant addition to the diets in Africa, southern Asia, and tropical America (FAO 2019b).

Fusarium oxysporum is a soil-borne fungus that is ranked fifth on the list of top fungal plant pathogens (Ploetz 2005; Dean et al. 2012). Over 120 *formae speciales* (ff. spp.) of *Fusarium oxysporum* have been described based on host specificity (Baayen et al. 2000). Differences in pathogenicity on specific host cultivars is being defined as physiological races among isolates (Kistler 1997; Baayen et al. 2000; Takken and Rep 2010; Meldrum et al. 2012). *Fusarium oxysporum* f. sp. *cubense* (Foc) refers to strains that infect bananas and plantains and cause Fusarium wilt or Panama disease (Ploetz 2005). It has been recognized that Foc has a polyphyletic origin (Lievens et al. 2009), hence comprises a suite of genetically distinct lineages (Ordonez et al. 2015). Maryani et al. (2019) have recently revised the taxonomy of Foc and designated different species names to strains affecting banana and merged them into the Fusarium of Banana Species Complex.

The disease cycle of this Fusarium spp. starts with infection of the root system and subsequent colonization within the vascular tissue, leads to water stress, severe chlorosis, and wilting (Ploetz 2015). Infected plants frequently die before they produce bunches, hence Fusarium wilt significantly reduces yields in infested fields (Dita et al. 2010).

A variant of Foc, called tropical race 4 (TR4) was first identified in Taiwan in 1989 but was probably the cause of banana wilt in the country since 1960. In the 1990s, Foc TR4 was identified in Malaysia and Indonesia, and the strains are thought to have originated from Taiwan (Buddenhagen 2009; Maryani et al. 2019). Foc TR4 has spread to many countries of Asia, as well as Australia and Africa and recently has reached Colombia and Peru in Latin America. Since its appearance, TR4 has severely affected Cavendish plantations in Malaysia, Indonesia, South China, Philippines, and the Northern Territory of Australia (Ploetz 2006; Molina et al. 2010; Buddenhagen 2009; Chittarath et al. 2018). TR4 is considered one of the most destructive Foc strains because of its broad host range. This pathogen is attacking the important cultivars of Cavendish but also all other cultivars that are sensitive to Foc (Cheng et al. 2019). The disease predominantly affects the Cavendish varieties, which not only primarily meets the international market demand but also is important for local consumption in developing countries. Cavendish varieties are the cornerstone for international trade, therefore TR4 threatens the entire global production (FAO 2019b).

Strategies controlling TR4 spread are based on visual monitoring of early symptom appearance, eradication of infected plants and isolation of infested areas to reduce pathogen dissemination. Pérez Vicente et al. (2014) reported that once plants are infected with TR4, there is no way to eradicate the disease. In this case the affected plants and all plants in the surrounding 7.5 m radius should be destroyed. Host resistance is a basis for sustainable disease management in most crops and this is usually achieved by intensive breeding programs (Ploetz 2006). Therefore, breeding for resistant/tolerant banana plants is the best way to overcome the disease.

To develop new, resistant cultivars, breeders need reliable and rapid phenotyping methods enabling selection of improved lines (García-Bastidas et al. 2019).

Different approaches can be pursued for resistance screening e.g. in the field or under greenhouse conditions (Dhingra and Sinclair 1986; Trigiano et al. 2004; Singh and Singh 2005). Field screening encounters problems such as time, costs, variable environmental conditions, and unspecified biodiversity of soil-borne pathogens (Mert and Karakaya 2003; Subramaniam et al. 2006; Sutanto et al. 2013). In contrast, greenhouse-based phenotyping facilitates high-throughput selection under controlled conditions with specific fungal genotypes, leading to more reproducible results (Smith et al. 2008). Greenhouse assessments have been reported as a reliable method by several researchers (Smith et al. 2008; Pérez Vicente et al. 2014). *In vitro* screening is one of the most high-throughput and efficient method (Švábová and Lebeda 2005; Pillay 2002; Naserian Khiabani et al. 2018; Wu et al. 2010). Compared to selection in an experimental field, *in vitro* selection can considerably reduce the space needed for screening. However, some factors influencing *in vitro* selection may differ from those in field selection (Matsumoto et al. 2010). The most used selection agents in the tissue culture medium are metabolites of pathogens or similar chemicals. There have been several reports of the use of fusaric acid to select Fusarium resistance in *in vitro* culture system (Matsumoto et al. 2010; Wu et al. 2010; Švábová and Lebeda 2005). Daub (1986) used a crude filtrate of a *Fusarium* suspension as a selection agent under the *in vitro* condition. Wu et al. (2010) and Naserian Khiabani et al. (2018) used suspensions containing pathogenic components including micro and macro-conidia and mycelium of *Fusarium oxysporum*, as a selection agent for *in vitro* screening. Using methionine sulfoximine as a selective agent, Carlson (1973) demonstrated for the first time the possibility of selecting disease-resistance plants using an *in vitro* tobacco protoplast system. Since 1980, the theoretical and practical approaches of *in vitro* selections and their usefulness for plant breeding have been addressed (Shepard 1981; Wenzel 1985; Daub 1986; Buiatti and Ingram 1991; Švábová and Lebeda 2005). According to Lebeda and Svábová (2010) the ideal system for *in vitro* selection for disease resistance should comprise: (1) an *in vitro* explant culture that can generate genetic variations (or an *in vitro* mutation induction system) with efficient recovery of genetically stable and fertile resistant/tolerant plants; (2) a selection agent that can be readily produced and which induces similar biochemical reactions *in vitro* as the pathogen *in vivo*; and, (3) molecular tools to characterize the selected resistant lines at the DNA level.

Several successful experiments have been carried out *in vitro* with live inoculums. For example: *Clavibacter michiganensis* (Bulk et al. 1991); *Xanthomonas campestris* (Hammerschlag 1990); *Mycosphaerella musicola* (Trujillo and Garcia 1996); *Alternaria alternata* (Takahashi et al. 1992); *Fusarium solani* (Huang and Hartman 1998); and *Phytophthora cinnamoni* (McComb et al. 1987; Cahill et al. 1992). Matsumoto et al. (2010) used fusaric acid as a selection agent in an *in vitro* culture system to select banana plants resistant to Fusarium wilt race 1.

Basic knowledge about the biology of the causal agent and its relationship with the host plant is necessary to develop suitable methods for resistance screening and selection (Russell 1978). Usually, wilting is either caused by blockage of plant vessels due to the accumulation of spores and mycelium of pathogenic fungi or due to a toxic element produced by the pathogen. In case of *Fusarium*, the use of

6. Pyridoxine hydrochloride.
7. Nicotinic acid.
8. Myo-inositol.
9. 6-Benzyl amino purine (BAP).
10. Sucrose.
11. Murashige and Skoog basal salt (MS) (Murashige and Skoog 1962).
12. Tissue culture grade water.
13. Gelling agent (e.g. Gelrite).
14. Tween 20.
15. Ethanol.
16. Sodium hydroxide (NaOH).
17. Sodium hypochlorite.
18. Magnetic stir bar.
19. Hot plate.
20. pH meter.
21. Erlenmeyer flasks.

2.2 Culture Media for the Isolation and Culture of *Fusarium oxysporum*

1. Petri plates.
2. 0.22 μm Cellulose Acetate (CA) filter.
3. Whatman filter (pore size 8 μm).
4. Potatoes.
5. Dextrose.
6. Agar.
7. Streptomycin (or other suitable antibiotic).
8. Hole-puncher.
9. Microscope.
10. Haemocytometer slide.

2.3 Biological Materials

1. Healthy banana suckers.
2. *Fusarium oxysporum* isolate.

2.4 Gamma Irradiation

1. Gamma Cell with ^{60}Co (here Gamma cell PX-30 was used).
2. High quality, disease-free *in vitro* banana plantlets.

3 Method

3.1 Preparation of Micropropagation and Rooting Medium

1. Prepare stock solutions (1000X or 1 mg/ml) of Thiamine, BAP, Nicotinic Acid, Pyridoxine (*see* Note 1).
2. Prepare stock solution of macro (10X concentrate) and micro (100X concentrate) elements of MS basal salt (MS medium).
3. For 1 l of the liquid micropropagation media, use the following amounts: 100 ml of MS macro salts, 10 ml of a MS micro salts, 5 ml BAP, 0.1 ml of Thiamine, 0.5 ml of Nicotinic acid and Pyridoxine solutions, 100 mg Myo-inositol, and 30 g sucrose. Dissolve in double distilled water.
4. For 1 l rooting medium use the following amounts: 50 ml of MS macro salts, 5 ml of a MS micro salts, 2 ml BAP and 0.1 ml of NAA, 0.1 ml of Thiamine, 0.5 ml of Nicotinic acid and Pyridoxine stock solutions, 100 mg Myo-inositol, 30 g sucrose, 7 g agar, and 3 g activated charcoal. Use double distilled water.
5. Place the media on the mixer and let it mix properly.
6. Calibrate the pH meter as per manufacturer instruction.
7. While stirring, adjust medium to pH 5.8 using NaOH or HCl.
8. After adjusting pH, top up the medium by adding the distilled water to 1 l.
9. Dispense 20 ml of the culture medium per jar.
10. Autoclave for 20 min at 121 °C.
11. Allow the medium to cool down.
12. Store the medium for up to a week in a cold room or in a refrigerator with 4–8 °C.

3.2 Preparation of Streptomycin Stock

1. Prepare Streptomycin (10 mg/ml) stock solution.
2. Filter Streptomycin solution using 0.22 μm cellulose acetate filters.
3. Keep the solution in a cold condition (4–5 °C).

3.3 Preparation of PDA (Potato Dextrose Agar) Medium

1. For 1 l of PDA, use 100 g of peeled potatoes (*see* Note 2).
2. Cut the peeled potatoes into four parts.
3. Boil potatoes for an hour in 200 ml of distilled water and filter through eight layers of cheesecloth.
4. Discard the solid portion; then add 10 g of dextrose and 6–7 g of agar to the clear liquid filtrate. Adjust the solution volume to 1 l by adding distilled water.

5. Dissolve well and heat the solution on a stirrer or in microwave (it usually takes approx. 40–50 min on the stirrer or 3–4 min in the microwave).
6. Dispense 500 ml in Erlenmeyer flasks (*see* Note 3) and autoclave immediately (100 kPa at 121 °C for 20 min).
7. Let it to cool to 50 °C.
8. Add 1.2 ml of 10 mg/ml Streptomycin solution.
9. Swirled the container by hand until the solution is homogenous.
10. Dispense 20 ml of the media in the Petri plates.

3.4 Preparation of Solid Inoculation Medium (SIM)

1. For 1 l of the solid culture media (1/2 MS), use the following amounts: 50 ml of Macro (10X), 5 ml of Micro (100X) solution, 7.5 g of sucrose, and 7 g of Agar. Use double distilled water (*see* Note 4).
2. Place the media on the mixer and let it mix properly.
3. Calibrate the pH meter as per manufacturer instruction.
4. While stirring, adjust medium to pH 5.8 using NaOH or HCl.
5. After adjusting pH, top up the medium by adding distilled water to 1 l.
6. Autoclave for 20 min at 120 °C.
7. Allow the medium to cool down.
8. Dispense 20 ml of the culture medium per jar.
9. Store the medium for up to 1 week in a cold room (keep in refrigerator with 4–5 °C).

3.5 Preparation of Inoculum

1. Inoculate the Foc single-spore isolates on Petri plates containing potato dextrose agar (PDA) (*see* Note 5).
2. Incubate for 1 week at 28 °C in the dark.
3. Cover the surface of fungal colonies by adding 5 ml of sterile water (Fig. 4.1).

Fig. 4.1 Preparation of the inoculum (**a**) *Fusarium oxysporum* colony (**b**) collection of conidia. (**c**) Counting cells in a haemocytometer (Source: http://insilico.ehu.eus/counting_chamber/thoma.php)

4. Gently probe the surface of the colony with the Pasteur pipette tip generating a mixture of conidial and hyphal fragments.
5. Pour the suspension into a sterile Falcon tube.
6. Count collected conidia (macro and micro-conidia) using a haemocytometer.
7. Adjust inoculum concentration to 10^6 conidia/ml.
8. Use the following formula to calculate the conidia density: $N \times 10^4 \times f$ cell/ml, where "N" is the total counted cells and "f" is dilution factor (*see* Note 6).

3.6 Preparation of Filter Paper Disks

1. Punch the Whatman filter paper with a hole-puncher with a 5 mm diameter.
2. Put discs into an autoclavable container.
3. Autoclave twice for 20 min or one time for 40 min at 120 °C (*see* Note 7).

3.7 Establishment of In Vitro Cultures

1. Collect small and healthy suckers from a healthy plant.
2. Wash suckers with tap water and cut into $10 \times 10 \times 10$ mm blocks containing shoot tips.
3. Surface-sterilize the shoot tips in a laminar flow cabinet with 70% Ethanol for 30 s, followed by 5% sodium hypochlorite with a few drops of Tween 20 for 30 min.
4. Transfer shoot tips into banana propagation liquid medium consisting of MS (Murashige and Skoog 1962) salts and vitamins, 5 mg/l BAP and 30 g/l sucrose.
5. Incubate shoot tips in a culture room (25 ± 1 °C, 16/8 day/night, 45–60 μmol/m^2/s light intensity).
6. Subculture every 30 to 45 days (*see* Note 8).

3.8 In Vitro Mutagenesis

1. Prepare at least 800 shoot tips from established *in vitro* banana plantlets.
2. Transfer shoot tips into sterile Petri dishes.
3. Add a few drops of sterile water and seal with parafilm.
4. Irradiate using a suitable dose (*see* Note 9).
5. Incubate shoot tips in liquid media for 45 days at 25 ± 1 °C and 16/8 day/night.
6. Subculture at least three times for chimera dissolving. After the first subculture, give each plantlet a unique code to keep track of the mutant pedigree.

7. Separate each M_1V_3 propagated individual into two parts. Place one part of plantlet from each shoot onto the same propagation media, and the second half place on the rooting media (*see* Fig. 4.2). These plantlets are M_1V_4 (*see* Note 11).
8. Select healthy and well rooted mutant plantlets and transfer into the SIM media.

Greenhouse plants

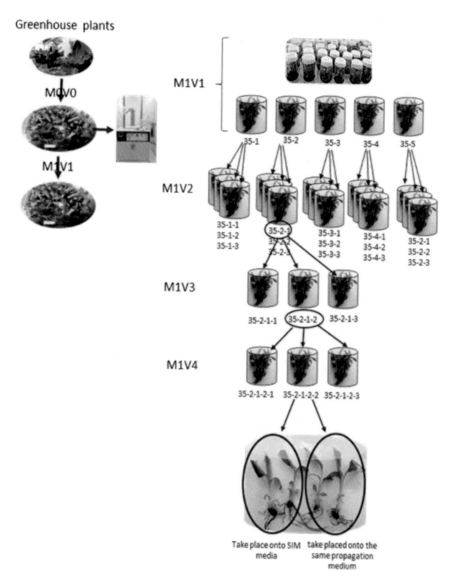

Fig. 4.2 Prepare the mutant population for *in vitro* bioassay by pedigree method, each genotype is evaluated for disease resistance, while the same genotype is being kept for subsequent studies, as well as for the reproduction of putative resistant mutants

3.9 In Vitro *Inoculation*

1. Select plantlets that meet the required criteria (*see* Fig. 4.3 and Note 12).
2. Incubate plantlets with healthy roots on SIM media for 2 weeks (*see* Note 13).
3. Soak 5 mm diameter filter paper discs in a conidial suspension.
4. Inoculate by putting one disc on the surface of the SIM media (*see* Fig. 4.4 and Note 15).
5. Incubate the inoculated plantlets at 25 ± 1 °C and 16/8 h day/night for 21–30 days.

3.10 *Disease Evaluation*

1. Perform daily observations of *Fusarium oxysporum* development on the SIM medium.
2. Assay disease according to symptoms appearing 1 week after inoculation (*see* Note 20).
3. Rate disease severity on a scale of 0 to 6 (*see* Table 4.1 and Fig. 4.5).
4. Record the symptoms and the rate of disease severity 21 to 30 days post-inoculation.
5. Select all plantlets that show resistance (rate 0–2) (*see* Note 21).

Fig. 4.3 Appropriate seedlings for *in vitro* evaluation of disease. Two-month-old seedlings 4–5 cm height of the pseudostem, have more than two fully expanded leaves and at least three white healthy roots

Fig. 4.4 Steps of inoculation: (**a**) selected plantlet incubated for 2 weeks on SIM media before inoculation, (**b**) autoclaved filter paper disks, (**c**) soaked filter disk in the inoculum suspension and placed on the surface of the SIM, (**d**) to (**f**) development of disease symptoms

Table 4.1 Disease severity-rating scale used to record symptoms caused by *Fusarium oxysporum* f.sp. *cubense* in banana plants (Wu et al. 2010)

Disease severity	Disease symptoms
0	Corm completely clean, no vascular discoloration
1	Smaller leaves at the base of pseudostem wilted, no discoloration of the pseudostem
2	≤1/2 the height of the pseudostem discolored
3	>1/2 the height of the pseudostem discolored and (or) there was discoloration of the leaf stalk
4	≤50% of the leaves wilted or yellowed
5	>50% of the leaves wilted or yellowed
6	The whole plantlet wilted

4   Notes

1. It is convenient to prepare the concentrated stock solutions of macro-salts, micro-salts, vitamins, amino acids, hormones, etc. All stock solutions should be stored in a refrigerator and should be checked visually for contamination with microorganisms or precipitation of ingredients. A stock solution of vitamins, amino acids, and hormones should not be stored for an indefinite period and should be kept −20 °C.

2. For preparing PDA, 100 g peeled potatoes are needed. It is therefore recommended to weigh them after peeling. The peeled potatoes are cut into four-parts. Each section should not be too small or crushed. Starch should not

Fig. 4.5 An example of *in vitro* bioassay of *Fusarium oxysporum*. M_1V_4 banana plantlets (local dwarf Cavendish CV) were used. The numbers (1 to 6) indicate disease severity scored according to Wu et al. (2010) after 21–30 days of inoculation

    enter into the final extract, and for the same reason, potatoes should not mash during boiling and filtering. A clear extract is needed for preparation of PDA.

3. To prevent liquids from spilling during the autoclaving process, dispense it into smaller volumes and pour it into larger containers. For example, pour 250 ml into a 500 ml container.

4. It is possible to utilize live pathogen for disease resistance screening in *in vitro* conditions. However, the *in vitro* conditions (higher humidity, reduced air velocity, media-rich in nutrients) favors the growth of microorganisms in general. To control the growth of the pathogen, the concentration of mineral salts, as well as the carbon source is reduced (*see* Sect. 3.4).

5. The protocol provided by Pérez Vicente et al. (2014) was used to isolate *F. oxysporum* and for single spore culture.

6. To count cells using a haemocytometer, add 15–20 µl of the cell suspension between the haemocytometer and a cover glass. Count the number of cells in all four outer squares, divide by four (*see* Fig. 4.1c, the cells in green will be counted). The number of cells per square $\times 10^4$ = the number of cells/ml of the suspension.

7. It is recommended to autoclave the filter paper twice or to increase the time of autoclaving to ensure the filter paper is sterile.

8. During each subculturing, abnormal or contaminated explants were removed.

9. The shoot tips were irradiated in a ^{60}Co gamma cell (dose rate was 0.018 Gy/s) with doses of 35 and 40 Gy. These doses were determined by the radiosensitivity test. After irradiation, the explants were placed onto a fresh multiplication medium. Each growing shoot was separately multiplied up to M_1V_3.

10. About 6000 individual plantlets were screened using the *in vitro* bioassay described here. Finally, 21–30 days after inoculation, a total of 50 putative mutant plantlets were selected with scores 0–2. These were categorized as putative resistant mutants.

11. Plants which have been selected in the process of *in vitro* screening must be kept for further evaluation. It is therefore necessary to back up the mutant plant at the screening stage by keeping a clone of M_1V_4 shoot tip of each tested plant. Chimera usually dissolves after three subcultures so that mutant genotypes belonging to the same progeny should be uniform after M_1V_4.

12. In the process of selecting seedlings resistant to Fusarium disease the test material used for evaluation should have the same size. It is recommended to select plantlets with pseudostem length ranging between 4.5 to 5 cm. Selected plantlets should have more than two fully expanded leaves and at least three white roots.

13. Two weeks before inoculation, uniform banana plantlets should be transferred to SIM media. Plantlets under the selection must be healthy. Any stress may affect the result of pathogen screening.

14. Young plantlets with small size are very susceptible to the disease. The best size of the plantlet is 4–5 cm, with 2–4 leaves and roots.

15. The paper disk should be put next to the plantlet. Putting the inoculation source near (less than 1 cm) to the plantlet leads the pathogen to reach the seedlings faster.

16. During the TR4 screening process care should be taken not to spread the disease. Therefore, all containers, media, and infected plantlets should be autoclaved after the termination of the bioassay.

17. While preparing the inoculum suspension, care should be taken that the solution does not get contaminated with bacteria. Adding antibiotics to the inoculum solution is a way to prevent such infection. Streptomycin or Chloramphenicol is suitable for this purpose.

18. This protocol was developed as a fast and cost-effective method for early mass screening. It enables reduction of mutant population and identification of putative mutants for further evaluation. The *in vitro* resistant plantlets should be evaluated under the field conditions for confirmation of their resistance/tolerance.

19. To enhance root formation, it is recommended that banana plantlets are transferred to the rooting medium. However even without this step, the plantlets with 4–5 cm height would normally produce sufficient roots.

20. Internal symptoms should be checked and scored in the case of plantlets with symptoms such as wilting or discoloration.

21. TR4 symptoms in the field and in *in vitro* conditions are distinct. In the field, wilting is the predominant disease symptom. By contrast, due to high humidity in the *in vitro* assay, wilting is less significant. Instead, the appearance of

necrotic leaves is more pronounced. Therefore, for disease scoring in the *in vitro* bioassay, one should especially score the appearance of necrotic and discolored leaves and pseudostem.

Acknowledgments I gratefully acknowledge the Joint FAO/IAEA Centre of Nuclear Techniques in Food and Agriculture Vienna, Austria, and Nuclear Science and Technology Research Institute, Karaj, Iran for their financial support. The author would like to acknowledge the Plant Breeding research group of Nuclear Agriculture Research School for their support and contribution to this study. I also thank Dr. Hamideh Afshar Manesh for assistance with the isolate *F. oxysporum* and Mr. Cyrus Vedadi for valuable discussions.

References

Baayen RP, O'Donnell K, Bonants PJ, Cigelnik E, Kroon LP, Roebroeck EJ, Waalwijk C (2000) Gene genealogies and AFLP analyses in the Fusarium oxysporum complex identify monophyletic and nonmonophyletic formae speciales causing wilt and rot disease. Phytopathology 90(8): 891–900. https://doi.org/10.1094/PHYTO.2000.90.8.891

Buddenhagen IW (2009) Understanding strain diversity in Fusarium oxysporum f. sp. cubense and history of introduction of "tropical race 4" to better manage banana production. Acta Hortic 828: 193–204

Buiatti M, Ingram D (1991) Phytotoxin as tools in breeding and selection of disease resistant plants. Experientia 47:811–819

Bulk R, Jansen J, Lindhout W, Löffler H (1991) Screening of tomato somaclones for resistance to bacterial canker Clavibacter michiganensis subsp. michiganensis. Plant Breed 107:190–196

Cahill D, Benett I, McComb A (1992) Resistance of micropropagated Eucalyptus marginata to Phytophthora cinnamomi. Plant Dis 76:630–632

Carlson P (1973) Methionine sulfoximine-resistant mutants of tobacco. Science 180:1366–1368

Cheng C, Liu F, Sun X, Tian N, Mensah RA, Li D, Lai Z (2019) Identification of Fusarium oxysporum f. sp. cubense tropical race 4 (Foc TR4) responsive miRNAs in banana root. Sci Rep 9(1):13682. https://doi.org/10.1038/s41598-019-50130-2

Chittarath K, Mostert D, Crew KS, Viljoen A, Kong G, Molina AB, Thomas JE (2018) First report of Fusarium oxysporum f. sp. cubense tropical race 4 (VCG 01213/16) associated with Cavendish bananas in Laos. Plant Dis 102(2):449–450

Daub M (1986) Tissue culture and the selection of resistance to pathogens. Annu Rev Phytopathol 24:159–186

Dean R, Van Kan JA, Pretorius ZA, Hammond-Kosack KE, Di Pietro A, Spanu PD, Rudd JJ, Dickman M, Kahmann R, Ellis J, Foster GD (2012) The top 10 fungal pathogens in molecular plant pathology. Mol Plant Pathol 13(4):414–430. https://doi.org/10.1111/j.1364-3703.2011. 00783.x

Dhingra OD, Sinclair JB (1986) Basic plant pathology methods. CRC Press, Boca Raton

Dita MA, Waalwijk C, Buddenhagen IW, MTJ S, GHJ K (2010) A molecular diagnostic for tropical race 4 of the banana Fusarium wilt pathogen. Plant Pathol 59:348–357

FAO (2019a) Banana market review preliminary results for 2019. Food and Agriculture Organization of the United Nations. http://www.fao.org/fileadmin/templates/est/COMM_MARKETS_MONITORING/Bananas/Documents/Banana_Market_Review_Prelim_Results_2018.pdf

FAO (2019b) Fusarium tropical race 4 (TR4). FAO. http://www.fao.org/world-banana-forum/fusariumtr4/en/

García-Bastidas FA, Van der Veen AJT, Nakasato-Tagami G, Meijer HJG, Arango-Isaza RE, Kema GHJ (2019) An improved phenotyping protocol for Panama disease in banana. Front Plant Sci 10:1006–1006. https://doi.org/10.3389/fpls.2019.01006

Hammerschlag FA (1990) Resistance response of plants regenerated from peach callus cultures to Xanthomonas campestris pv. pruni. J Am Soc Hortic Sci 115:1034–1037

Huang Y, Hartman G (1998) Reaction of selected soybean genotypes to isolates of Fusarium solani f. sp. glycines and their culture filtrates. Plant Dis 82(9):999–1002

Jain SM (2000) Mechanisms of spontaneous and induced mutations in plants. In: 11th international congress of radiation research, Dublin, Ireland, 18–23, July 1999, pp 255–258

Jain SM (2006) Mutation assisted breeding in ornamental plant improvement. Acta Hortic 714:85–98

Jain SM (2007) Recent advances in plant tissue culture and mutagenesis. Acta Hortic 736:205–211

Jain SM (2010). In vitro mutagenesis in banana (Musa spp.) improvement. In: T Dubois et al. (ed) IC on banana & plantain in Africa, 2010, vol 879, ISHS. Acta Hortic

Jain SM, Maluszynski M (2004) Induced mutations and biotechnology in improving crops. In vitro application in crop improvement. Science Publishers, London

Kistler HC (1997) Genetic diversity in the plant-pathogenic fungus fusarium oxysporum. Phytopathology 87(4):474–479. https://doi.org/10.1094/PHYTO.1997.87.4.474

Lebeda A, Svábová L (2010) In vitro screening methods for assessing plant disease resistance. In: IAEA (ed) Mass screening techniques for selecting crops resistant to diseases, chap. 2, joint FAO/IAEA programme of nuclear techniques in food and agriculture. International Atomic Energy Agency, Vienna, pp 5–46

Lievens B, Houterman PM, Rep M (2009) Effector gene screening allows unambiguous identification of Fusarium oxysporum f. sp. lycopersici races and discrimination from other formae speciales. FEMS Microbiol Lett 300(2):201–215. https://doi.org/10.1111/j.1574-6968.2009.01783.x

Maryani N, Lombard L, Poerba YS, Subandiyah S, Crous PW, Kema GHJ (2019) Phylogeny and genetic diversity of the banana Fusarium wilt pathogen Fusarium oxysporum f. sp. cubense in the Indonesian centre of origin. Stud Mycol 92:155–194. https://doi.org/10.1016/j.simyco.2018.06.003

Matsumoto K, Barbosa M, Souza L, Teixeira J (2010) In vitro selection for resistance to Fusarium wilt in Banana. In: Mass screening techniques for selecting crops resistant to diseases. International Atomic Energy (IAEA), Vienna, pp 101–114

McComb JA, Hinch JM, Clarke AE (1987) Expression of field resistance in callus tissue inoculated with Phytophthora cinnamoni. Phytopathology 46:346–351

Meldrum RA, Fraser-Smith S, Tran-Nguyen LTT, Daly AM, Aitken EAB (2012) Presence of putative pathogenicity genes in isolates of Fusarium oxysporum f. sp. cubense from Australia. Australas Plant Pathol 41:551–557

Mert Z, Karakaya A (2003) Determination of the suitable inoculum concentration for Rhynchosporium secalis seedling assays. J Phytopathol 151:699–701

Molina AB, Williams RC, Hermanto C, Suwanda B, Komolong B, Kokoa P (2010) Final report: mitigating the threat of banana Fusarium wilt: understanding the agro ecological distribution of pathogenic forms and developing disease management strategies. ACIAR Publication, Canberra

Murashige T, Skoog F (1962) A revised medium for rapid growth and bioassays with tobacco tissue cultures. Physiol Plant 15:473–497

Naserian Khiabani B, Vedadi C, Afsharmanesh H, Eskandari A (2018) In-vitro mutation breeding and selection for resistance to fusarium wilt in Banana. Paper presented at the International symposium on plant mutation breeding and biotechnology, Vienna, Austria, 27–31 August

Novák FJ, Afza R, Van Duren M, Perea-Dallos M, Conger BV, Xiolang T (1989) Somatic embryogenesis and plant regeneration in suspension cultures of dessert (AA and AAA) and cooking (ABB) bananas (Musa spp.). Biotechnology 46:125–135

Ordonez N, Seidl MF, Waalwijk C, Drenth A, Kilian A, Thomma BP, Ploetz RC, Kema GH (2015) Worse comes to worst: bananas and Panama disease – when plant and pathogen clones meet. PLoS Pathog 11(11):e1005197. https://doi.org/10.1371/journal.ppat.1005197

Pérez Vicente L, Dita M, Martinez De La Parte E (2014) Technical manual: prevention and diagnostic of fusarium wilt (Panama disease) of banana caused by fusarium oxysporum f. sp. cubense tropical race 4 (TR4). FAO, Rome

Pillay M (2002) Future challenges in Musa breeding. In: Crop improvement for the 21st century. Routledge, London

Pillay M, Tenkouano A (2011) Genomes, cytogenetics and flow cytometry of Musa. In: Banana breeding progress and challenges. CRC Press, Boca Raton

Ploetz RC (2005) Panama disease: an old nemesis rears its ugly head: part 1. The beginnings of the banana export trades. Plant. Health Prog 6(1):18

Ploetz R (2006) Fusarium wilt of banana is caused by several pathogens referred to as Fusarium oxysporum f. sp. cubense. Phytopathology 96:653–656

Ploetz R (2015) Management of Fusarium wilt of banana: a review with special reference to tropical race 4. Crop Prot 73:7–15

Predieri S (2001) Mutation induction and tissue culture in improving fruits. Plant Cell Tissue Organ Cult 64:185–210

Roux NS (2004) Mutation induction in Musa-review. In: Banana improvement: cellular, molecular biology and induced mutations. Science Publishers, Enfield

Russell GE (1978) Plant breeding for Pest and disease resistance. Butterworths, London

Shepard J (1981) Protoplasts as sources of disease resistance in plants. Annu Rev Phytopathol 19: 145–166

Singh DP, Singh A (2005) Disease and insect resistance in plants. Science Publishers, Enfield

Smith LJ, Smith MK, Tree D, O'keefe D, Galea VJ (2008) Development of a small-plant bioassay to assess banana grown from tissue culture for consistent infection by Fusarium oxysporum f. sp. cubense. Australas Plant Pathol 37:171–179

Subramaniam S, Maziah M, Sariah M, Puad M, Xavier R (2006) Bioassay method for testing Fusarium wilt disease tolerance in transgenic banana. Sci Hortic 108:378–389

Sutanto A, Sukma D, Hermanto C (2013) The study and early evaluation of resistance of banana accessions for wilt disease caused by Fusarium oxyporum f. sp. cubense VCG 01213/16 (TR4). In: Improving food, energy and environment with better crops. 7th Asian crop science association conference, IPB international convention Center, Bogor, Indonesia, 27–30 September 2011. Research Center for Bioresources and Biotechnology, Bogor Agricultural University, pp 291–295

Švábová L, Lebeda A (2005) In vitro selection for improved plant resistance to toxin-producing pathogens. J Phytopathol 153:52–64

Takahashi H, Takatsugu T, Tsutomu M (1992) Gene analysis of mutant resistant to Alternaria alternata strawberry pathotype selected from calliclones of strawberry cultivar Morioka-16. J Jpn Soc Hortic Sci 61:347–351

Takken F, Rep M (2010) The arms race between tomato and Fusarium oxysporum. Mol Plant Pathol 11:309–314

Trigiano RN, Windham MT, Windham AS (2004) Plant pathology concepts and laboratory exercises. CRC Press, Boca Raton

Trujillo I, Garcia DE (1996) Aplicación de metódos de presión de selección en la obtención de variantes de banano resistentes a la Sigatoka Amarilla. Phyton Int J Exp Bot 59:111–121

Van Harten AM (1998) Mutation breeding: theory and practical applications. Cambridge University Press, Cambridge

Wenzel G (1985) Strategy in unconventional breeding for disease resistance. Annu Rev Phytopathol 23:149–172

Wu Y, Yi G, Peng X (2010) Rapid screening of Musa species for resistance to Fusarium wilt in an in vitro bioassay. Eur J Plant Pathol 128(3):409–415

Chapter 5
An Optimised Greenhouse Protocol for Screening Banana Plants for Fusarium Wilt Resistance

Privat Ndayihanzamaso, Sheryl Bothma, Diane Mostert, George Mahuku, and Altus Viljoen

Abstract Fusarium wilt, caused by the soil-borne fungus *Fusarium oxysporum* f. sp. *cubense* (Foc), is considered one of the most devastating diseases of banana in the world. Effective management of Fusarium wilt is only achieved by planting banana varieties resistant to Foc. Resistant bananas, however, require many years of breeding and field-testing under multiple geographical conditions. Field evaluation is reliable but time consuming and expensive. Small plant screening methods are, therefore, needed to speed up the evaluation of banana varieties for Foc resistance. To this end, a small plant screening method for resistance to banana Fusarium wilt is presented. The method proposes the planting of 2- to 3-month-old banana plants in soil amended with 10 g Foc-colonised millet seeds. Rhizome discoloration is then evaluated to rank the disease resistance response. The optimized millet seed technique could be useful in mass screening of newly developed genotypes for resistance to Foc.

Keywords *Musa* · *Fusarium oxysporum* f. sp. *cubense* · Evaluation · Resistant varieties

1 Introduction

Fusarium wilt is considered one of the most devastating diseases of banana in the world (Stover 1962; Ploetz 2006). The disease is caused by a soil-borne fungus, *Fusarium oxysporum* f. sp. *cubense* (Foc), which infects the plants through the roots, colonises the rhizome and xylem, and causes a lethal wilting of plants (Stover 1962;

P. Ndayihanzamaso · S. Bothma · D. Mostert · A. Viljoen (✉)
Department of Plant Pathology, Private Bag X1, Stellenbosch University, Matieland, South Africa
e-mail: altus@sun.ac.za

G. Mahuku
International Institute of Tropical Agriculture (IITA) Regional Hub, Dar-es-Salaam, Tanzania

© The Author(s) 2022
J. Jankowicz-Cieslak, I. L. Ingelbrecht (eds.), *Efficient Screening Techniques to Identify Mutants with TR4 Resistance in Banana*,
https://doi.org/10.1007/978-3-662-64915-2_5

65

Guo et al. 2015; Warman and Aitken 2018). Foc originated in Southeast Asia, and was disseminated globally with Foc-propagating and planting material as banana production expanded during the twentieth century (Stover 1962; Ploetz 2015a; Dita et al. 2018). By the 1950s, Fusarium wilt was so widespread in Latin America that it became impossible to sustain the international banana export industry that was almost exclusively based on the Gros Michel (AAA) bananas. Gros Michel was thus replaced with Cavendish bananas as export variety in the 1960s, which currently account for nearly 50% of production globally (Lescot 2015; Ploetz 2015a). Cavendish bananas, however, are now also severely affected by a different strain of Foc in Asia, the Middle East, Mozambique in Africa and Colombia in Latin America, called Foc Tropical Race 4 (TR4) (Ploetz and Pegg 2000; Mostert et al. 2017; Dita et al. 2018; García-Bastidas et al. 2019a, b; Thangavelu et al. 2019; Viljoen et al. 2020).

Foc comprises of three races based on their pathogenicity to a group of differential cultivars, with Foc races 1, 2 and 4 causing disease to Gros Michel, Bluggoe and Cavendish bananas, respectively (Pegg and Langdon 1987; Ploetz and Pegg 2000). Foc race 4 is further subdivided into Foc subtropical race 4 (Foc STR4) and tropical race 4 (Foc TR4) strains. The former causes disease on Cavendish bananas in the subtropics when plants are stressed by adverse climatic conditions, while Foc TR4 affects Cavendish bananas both under tropical and subtropical conditions (Ploetz and Pegg 2000). Foc TR4 has a wider host range than Foc races 1 and 2, and causes disease to Cavendish banana cultivars as well as most Foc race 1- and race 2-susceptible cultivars (Ploetz 2015b). Foc strains have been further classified using vegetative compatibility group (VCG) analysis, which is based on the ability of fungal hyphae to anastomose and form a stable heterokaryon (Leslie and Summerell 2006). Twenty-four VCGs have been described for Foc (Pegg et al. 1993; Bentley et al. 1995; Ploetz and Correll 1998; Fourie et al. 2009).

The most effective method to control banana Fusarium wilt is the use of Foc-resistant varieties (Ploetz 2015b). If existing varieties with resistance are not available to replace susceptible ones, the susceptible bananas can be improved by conventional breeding, mutation breeding, genetic engineering and genome editing (Rowe 1987; Hwang and Tang 2000; Hwang and Ko 2004; Bakry et al. 2009; Ortiz 2013; Dale et al. 2017; Naim et al. 2018). Bananas resistant to Foc race 1 that have been developed by conventional breeding include the hybrids FHIA-01, FHIA-18 and FHIA-25 developed in Honduras; RG1, Bodles Altafort, 2390-2 and 72-1242 bred in Jamaica; Pacovan Ken, Preciosa and Tropical in Brazil; and BRS-01 and BRS-02 in India (Bakry et al. 2009; Lorenzen et al. 2012; Ortiz 2013). Several partially resistant and/or resistant clones have also been developed by mutation breeding through chemical and radiation mutagenesis techniques and through somaclonal variation. For instance, GCTCV-215, GCTCV-105, GCTCV-218 and GCTCV-219 are Giant-Cavendish derived somaclones, which were selected through somaclonal variation, and Dwarf Parfitt mutant DPM-25 is a banana mutant generated with gamma irradiation (Tang and Hwang 1994; Hwang and Ko 2004; Smith et al. 2006). Dale et al. (2017) developed two transgenic Cavendish lines, RGA2-3 and Ced9-21, which did not develop any Fusarium wilt after 3 years of evaluation in banana fields infested with Foc TR4. Apart from GCTCV-218, most improved bananas with Fusarium wilt resistance have, however, not satisfied export market requirements (Bakry et al. 2009; Ortiz 2013).

Before local varieties or improved bananas can replace susceptible bananas in Foc-infested fields, they need to be evaluated for resistance against local Foc strains (Morpurgo et al. 1994a; Ploetz 1994; Carlier and De Waele 2002; Mak et al. 2004; Dita et al. 2011). Field evaluation is more accurate when identifying Foc-resistant plants under natural environmental conditions, but the process is time-consuming and expensive. Inoculum levels in the soil might also be unequally distributed. In addition, banana varieties can only be tested against Foc strains present in the country or region where the tests are performed. Greenhouse screening can be performed on small plants, achieved in a short time, and can be screened against quarantine pathogens in restricted environments (Smith et al. 2008; Dita et al. 2011). Yet, greenhouse screening methods seldom correlate well with field screening due to a number of factors, including inoculum preparation, inoculum concentration, inoculation method, the effect of temperature and photoperiod, type of planting material tested, and the potting soil used (Brake et al. 1995; Smith et al. 2008; Dita et al. 2011). Greenhouse screening methods, thus, need optimisation.

Several greenhouse evaluation methods have been developed for banana Fusarium wilt (Morpurgo et al. 1994b; Brake et al. 1995; De Ascensao and Dubery 2000; Mohamed et al. 2001; Smith et al. 2008; Wu et al. 2010; Dita et al. 2011; Viljoen et al. 2018; Chen et al. 2019; García-Bastidas et al. 2019a, b). The most common inoculation methods include the dipping of plant roots in a conidial suspension, the drenching of soil with a conidial suspension, and the replanting of plants in Foc-infested soil or sand. In the dipping method, plants are carefully removed from soil and their roots dipped in a conidial suspension for a few minutes before replanting (Mohamed et al. 2001; Dita et al. 2011; Ribeiro et al. 2011). The soil-drenching method consists of pouring a spore suspension on the surface of potting soil (Smith et al. 2008). In the infested soil technique, bananas are planted in soil mixed with millet seeds or maize kernels that were pre-colonized with Foc (Smith et al. 2008; Dita et al. 2011). A combination of the dipping method and replanting in infested soil was reported to result in quick and consistent disease development (Dita et al. 2011). The screening of banana plants grown *in vitro* has also been reported, but Hamill (2018) considered these not suitable for resistance screening, as the fungus may kill both susceptible and resistant varieties due to high inoculum pressure.

The type of planting material used can influence Fusarium wilt development in the greenhouse. Tissue culture-derived plants are more susceptible and have a shorter incubation period compared to suckers and bits (Hwang and Ko 1987; Smith et al. 2008). Tissue culture plants are also free of other pests and diseases, apart from plant viruses, which may influence disease development. Tests with tissue culture plants smaller than 5 cm did not reflect field results, but plants of 10–15 cm (around 2 months old) did (Brake et al. 1995; Mohamed et al. 2001; Smith et al. 2008). Further investigations of the effect of plant age on Fusarium wilt development are thus needed to improve the reliability of small plantlet screening methods (Brake et al. 1995; Smith et al. 2008; Ribeiro et al. 2011).

In this chapter, we present an optimised greenhouse screening method for resistance to banana Fusarium wilt, considering the effect of inoculum concentration, inoculation methods and plant age on disease development.

2 Materials

2.1 Preparation of Plant Material

1. Potting bags (3–5-l capacity).
2. Commercial potting soil (mix of screened compost, coarse sand and screened bark).
3. Building sand.
4. Coco peat.
5. Slow-release fertiliser.
6. Sporekill® (120 g/L didecyldimethyl-ammonium chloride) solution (ICA International Chemicals (PTY) Ltd., Stellenbosch, South Africa).
7. Seedling trays if plants are to be hardened-off.

2.2 Culture Medium

1. Potato dextrose agar (PDA) powder.
2. Deionised water.
3. Petri dishes (90-mm-diameter).
4. Analytical balance.
5. Weighing trays.
6. Spatula.
7. 1l Schott bottles.
8. Autoclave.

2.3 Preparation of Inoculum and Greenhouse Infection

1. Millet seed (*Panicum miliaceum* or *Eleusine corocana*).
2. 1l Schott bottle or 250 ml Erlenmeyer flasks.
3. Sterile distilled water.
4. Autoclave.

5. Scalpel.
6. Culture media (PDA).
7. *Fusarium oxysporum* f. sp. *cubense* isolate characterised to VCG level. Foc isolates can be obtained from the Westerdijk Institute in the Netherlands (https://wi.knaw.nl/page/Collection) and the Agricultural Research Service of the United States Department of Agriculture (https://nrrl.ncaur.usda.gov/).
8. Incubator with light and temperature control (light regime 65 μmol/m^2/s; e.g. cool-white fluorescent tubes, Philips TLP 36/86; temperature regime of 25 °C +/− 2 °C).
9. Analytical balance.
10. Greenhouse (temperature of 25 °C +/− 2 °C and relative humidity of 80%).

2.4 Disease Rating

1. Sterile tissue paper.
2. Ethanol for surface and tools sterilisation.
3. Forceps.
4. Scalpels.
5. Scalpel blades.
6. Knives.
7. 15 ml Falcon tubes.

3 Methods

3.1 Planting Material

1. Banana accessions to be tested are prepared together with resistant and susceptible control plants. Resistant and susceptible control varieties are selected according to the Foc strain to be used for the inoculation (Table 5.1) (Viljoen et al. 2017).

Table 5.1 Banana varieties used as resistant and susceptible checks for evaluation against races of *Fusarium oxysporum* f. sp. *cubense*

Pathogen	Susceptible control	Resistant control
Foc race 1	Gros Michel, Silk, Pisang Awak	Cavendish, Calcutta-4, Mbwazirume (EAHB[a])
Foc race 2	Bluggoe, Silk	Cavendish, Calcutta-4, Mbwazirume
Foc race 4	Cavendish, Silk, Gros Michel	Calcutta-4, Mbwazirume

[a]*EAHB* East African Highland Bananas

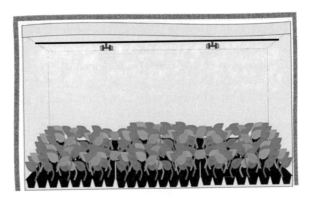

Fig. 5.1 Construction of a humidity chamber used for the weaning of tissue culture plants. The humidity chamber is constructed out of a metal framework covered in clear polyethylene sheeting. The humidity chamber is covered with shade netting and fitted with an automatic misting system. Seedling trays are placed on a lifted metal mesh to ensure the drainage of excess water

2. Nine to 15 plants per accession are needed for resistance screening. Additional susceptible control plants need to be included in the experiment to monitor the development of internal symptoms.
3. Plants must be produced in tissue culture, and the rooted plants weaned for 4–6 weeks in a humidity chamber (*see* Fig. 5.1). Light conditions are reduced to 20–30% transmission, and humidity increased to 60–90%.
4. Roots of the tissue cultures plants are gently washed with a Sporekill solution (1 ml Sporekill/l water) to remove all adhering media and to prevent opportunistic infections.
5. Excess leaf and root material is trimmed back, leaving only two to four soft green leaves and several white primary roots (*see* Fig. 5.2a). The plants are then replanted in plastic seedling trays filled with semi-sterile coco peat amended with a slow-release fertilizer (Osmocote start, 11:11:17 + 2 MgO + TE, 2 g/kg coco peat). The misting system is set to run for 15–20 s every 30 min in the warmer months, and 10 s every 30 min during cooler months.
6. Once the weaned plants have reached a height of 5–12 cm (*see* Fig. 5.2b), they are replanted into planting bags or pots containing a semi-sterile potting mix (commercial soil:sand:peat at 1:1:1) amended with slow release fertiliser (Osmocote Pro, 19:9:10 + 2MgO + TE, 2 g/kg soil mix) to support growth and development. The plants are kept on tables or raised structures in the greenhouse, and irrigated twice daily with an overhead micro-sprinkler system.
7. Plants in the hardening-off phase are ready to be screened after a period of 6–12 weeks, or once the plants have reached a height of 25–30 cm (*see* Fig. 5.2c).

Fig. 5.2 Plants during the weaning and hardening phases. (**a**) Tissue culture plant ready for weaning. (**b**) A plant at the end of the weaning phase that is ready to be hardened-off. (**c**) Hardened-off plants ready to be screened for Fusarium wilt resistance

3.2 Preparation of Culture Media

1. Add 39 g of PDA powder to 1l Schott bottle and fill up with deionised water.
2. Autoclave at 121 °C for 20 min.
3. Cool down to 50 °C.
4. Poor into 90-mm sterile Petri dishes.
5. Store PDA plates at 4 °C until use.

3.3 Millet Seed Inoculum Preparation

1. Fill 1l Erlenmeyer flasks or Schott bottles with 250 g millet seeds.
2. Add 200 ml of distilled water to soak the millet seeds overnight.
3. Drain-off all excess water from the millet seeds the following morning.
4. Autoclave the millet seed at 121 °C for 20 min on two consecutive days.

Fig. 5.3 Preparation of *Fusarium oxysporum* f. sp. *cubense* (Foc) inoculum on millet seed. (**a**) Autoclaved millet seeds ready for inoculation, (**b**) white fungal growth on the surface of the millet seeds. At this time, the flasks/Schott bottles need to be shaken to ensure proper distribution and thorough colonisation of millet seeds, and (**c**) the plating out of Foc-colonized millet seeds onto PDA plates to ensure proper colonization with the Foc inoculum, and no contamination with other micro-organisms

5. Shake the flask/bottle to loosen the grain (*see* Fig. 5.3a). A small sample is collected for plating onto culture media to ensure that it is sterile.
6. Plate out a Foc isolate on PDA and incubate it at 25 °C for 5–7 days.
7. Cut 10 mycelial plugs of 0.5 cm in diameter from the margins of the culture, and transfer these to the flasks/Schott bottles containing sterilized millet seeds.
8. When white fungal growth begins to show on the surface of millet seeds, the flasks/bottles are shaken to distribute the fungus and to prevent the kernels from clumping together (*see* Fig. 5.3b).
9. Let the fungus colonise the millet seed for 14 days, but shake every second day to ensure thorough colonisation of the millet seed. Plate out some seeds onto PDA to ensure that it is colonised by Foc only (*see* Fig. 5.3c). The millet seeds can then be stored at room temperature until use, but not for longer than 3 months to ensure the fitness and pathogenicity of Foc.

3.4   Greenhouse Inoculation

1. Fill planting bags/pots halfway with steam-sterilized potting soil, and mix with the Foc inoculum at a rate of 10 g millet seeds/kg soil (*see* Fig. 5.4a, b).
2. Uproot the plant (2-month-old) that needs to be inoculated carefully, and replant it into potting bags/pots containing the Foc-inoculated potting soil (*see* Fig. 5.4c). Then fill up the entire bag/pot with potting soil. Clearly mark the bags/pots with the accession name.
3. Once all plants are inoculated, they need to be randomly arranged in the greenhouse (*see* Fig. 5.4d).

Fig. 5.4 Inoculation of banana plants for resistance screening against *Fusarium oxysporum* f. sp. *cubense* (Foc). (**a**) Weighing-off of millet seed inoculum, (**b**) mixing Foc-millet seeds with potting soil, (**c**) uprooted plantlet ready for replanting in bags/pots containing Foc-infested soil, (**d**) inoculated plants set up in a screen house and (**e**) plants ready for disease development rating

4. Experimental conditions should include a 12 h daylight photoperiod at 25/20 °C, and 80% humidity.
5. Add 1 g of slow-release fertilizer per pot for the screening period, and irrigate plants to prevent any environmental stresses.

3.5 Scoring Disease Severity

1. Assess the host plant response when leaf symptoms are visible on 50% of the susceptible control plants, or when susceptible control plants have an internal disease rating of 5 and more. This is usually 6–12 weeks after inoculation.
2. For internal disease rating, the rhizome should be cut open horizontally in the middle of the rhizome, where discoloration is more pronounced than in the lower or upper rhizome (*see* Fig. 5.5a, b).
3. Score the rhizome discoloration on a rating scale ranging from 1 to 6, with 1 indicating no disease symptoms and 6 indicating complete discolored of the inner rhizome (*see* Fig. 5.5c).

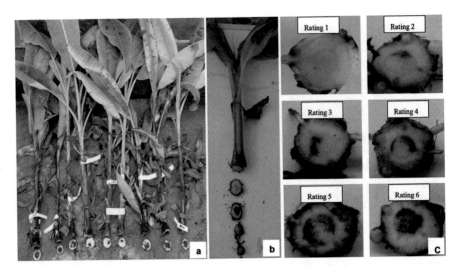

Fig. 5.5 Evaluation of Fusarium wilt development. (**a**) Rhizomes of inoculated plants are cut open for rating, (**b**) the position where the cut is made, and (**c**) rating is based on the discoloration of the inner rhizome as follows: 1 = no internal symptoms, 2 = few internal spots, 3 = <1/3 of the inner rhizome affected, 4 = 1/3–2/3 of the inner rhizome discolored 5 = >2/3 of the inner rhizome discolored and 6 = entire inner rhizome discolored

Acknowledgments The authors would like to acknowledge the University of Stellenbosch, International Institute of Tropical Agriculture (IITA) and the Joint FAO/IAEA Centre of Nuclear Techniques in Food and Agriculture for financial assistance.

References

Bakry F, Carreel F, Jenny C, Horry JP (2009) Genetic improvement of banana. In: Jain JM, Priyadarshan PM (eds) Breeding plantation tree crops: tropical species. Springer, New York, pp 3–50

Bentley S, Pegg KG, Dale JL (1995) Genetic variation among a worldwide collection of isolates of *Fusarium oxysporum* f. sp. *cubense* analysed by RAPD-PCR fingerprinting. Mycol Res 99: 1378–1384

Brake VM, Pegg AKG, Irwin JAG, Chaseling J (1995) The influence of temperature, inoculum level and race of *Fusarium oxysporum* f. sp. *cubense* on the disease reaction of banana cv. Cavendish. Aust J Agric Res 46:673–685

Carlier J, De Waele D (2002) Global evaluation of Musa germplasm for resistance to Fusarium wilt, Mycosphaerella leaf spot diseases and nematodes, INIBAP technical guidelines. INIBAP, Montpellier, p 57

Chen A, Sun J, Matthews A, Armas-Egas L, Chen N, Hamill S, Mintoff S, Tran-Nguyen LTT, Batley J, Aitken EAB (2019) Assessing variations in host resistance to *Fusarium oxysporum* f sp. *cubense* race 4 in *Musa* species, with a focus on the subtropical race 4. Front Microbiol 10: 1062. https://doi.org/10.3389/fmicb.2019.01062

Dale J, Anthony J, Paul JY, Khanna H, Smith M, Peraza-Echeverria S, Garcia-Bastidas F, Kema GHJ, Waterhouse P, Mengersen K, Harding R (2017) Transgenic Cavendish bananas with resistance to Fusarium wilt tropical race 4. Nat Commun 8:1496. https://doi.org/10.1038/s41467-017-01670-6

De Ascensao ARDCF, Dubery IA (2000) Panama disease: cell wall reinforcement in banana roots in response to elicitors from *Fusarium oxysporum* f. sp. *cubense* race 4. Phytopathology 90:1173–1180

Dita MA, Waalwijk C, Paiva LV, Souza MT, Kema GHJ (2011) A greenhouse bioassay for the *Fusarium oxysporum* f. sp. *cubense* x grand Naine (*Musa*, AAA, Cavendish subgroup) interaction. Acta Hortic 897:1–5

Dita MA, Barquero M, Heck D, Mizubuti ESG, Staver CP (2018) Fusarium wilt of banana: current knowledge on epidemiology and research needs toward sustainable disease management. Front Plant Sci 9:1468. https://doi.org/10.3389/fpls.2018.01468

Fourie G, Steenkamp ET, Gordon TR, Viljoen A (2009) Evolutionary relationships among the *Fusarium oxysporum* f. sp. *cubense* vegetative compatibility groups. Appl Environ Microbiol 75:4770–4781

García-Bastidas FA, Quintero-Vargas JC, Ayala-Vasquez M, Schermer T, Seidl MF, Santos-Paiva-M, Noguera AM, Aguilera-Galvez C, Wittenberg A, Hofstede R, Sorensen A, Kema GHJ (2019a) First report of Fusarium wilt tropical race 4 in Cavendish bananas caused by *Fusarium odoratissimum* in Colombia. Online publication: https://doi.org/10.1094/PDIS-09-19-1922-PDN. 3 Oct 2019

García-Bastidas FA, Van der Veen AJT, Nakasato-Tagami G, Meijer HJG, Arango-Isaza RE, Kema GHJ (2019b) An improved phenotyping protocol for Panama disease in banana. Front Plant Sci 10:1006. https://doi.org/10.3389/fpls.2019.01006

Guo L, Yang L, Liang C, Wang G, Dai Q, Huang J (2015) Differential colonization patterns of bananas (*Musa* spp.) by physiological race 1 and race 4 isolates of *fusarium oxysporum* f. sp. *cubense*. J Phytopathol 163:807–817

Hamill SD (2018) Rapid progression of disease in susceptible and resistant banana cultivars inoculated with *Fusarium oxysporum* f. sp. *cubense* race 1 and subtropical race 4 in tissue culture. Acta Hortic 1205:749–756

Hwang SC, Ko WH (1987) Somaclonal variation of bananas and screening for resistance to Fusarium wilt. In: Persley GJ, De Langhe EA (eds) Banana and plantain breeding strategies. ACIAR, Cairns, pp 151–156

Hwang SC, Ko WH (2004) Cavendish banana cultivars resistant to Fusarium wilt acquired through somaclonal variation in Taiwan. Plant Dis 88:580–588

Hwang SH, Tang CY (2000) Unconventional banana breeding in Taiwan. In: Jones DR (ed) Diseases of Banana, Abacá and Enset. CABI Publishing, Wallingford, pp 449–464

Lescot T (2015) Genetic diversity of the banana. FruitTrop 231:98–102

Leslie JF, Summerell BA (2006) The Fusarium laboratory manual, 1st edn. Blackwell Publishing, Iowa

Lorenzen J, Tenkouano A, Bandyopadhyay R, Vroh-Bi I, Coyne D, Tripathi L (2012) Overview of banana and plantain improvement in Africa: past and future. Acta Hortic 879:595–604

Mak C, Mohamed AA, Liew KW, Ho YW (2004) Early screening technique for Fusarium wilt resistance in banana micropropagated plants. In: Swennen R, Jain MS (eds) Banana improvement: cellular, molecular biology, and induced mutations. Science Publishers, New Hampshire, pp 219–227

Mohamed AA, Mak C, Liew KW, Ho YW (2001) Early evaluation of banana plants at nursery stage for fusarium wilt tolerance. In: Molina AB, Masdek NH, Liew KW (eds) Banana Fusarium wilt: towards sustainable cultivation. INIBAP-ASPNET, Los Banos, pp 174–185

Morpurgo R, Lopato SV, Afza R, Novak FJ (1994a) Selection parameters for resistance to *Fusarium oxysporum* f. sp. *cubense* race 1 and race 4 on diploid banana (*Musa acuminata* Colla). Euphytica 75:121–129

Morpurgo R, Duren MVAN, Grasso G, Afza R (1994b) Differential response of banana cultivars to *Fusarium oxysporum* f. sp. *cubense* infection for chitinases activity. In: Mass screening techniques for selecting crops resistant to disease. IAEA, Vienna, pp 129–134

Mostert D, Molina AB, Daniells J, Fourie G, Hermanto C, Chao CP, Fabregar E, Sinohin VG, Masdek N, Thangavelu R, Li C, Yi G, Mostert L, Viljoen A (2017) The distribution and host range of the banana Fusarium wilt fungus, *Fusarium oxysporum* f. sp. *cubense*, in Asia. PLoS One 12:e0181630. https://doi.org/10.1371/journal.pone.0181630

Naim F, Dugdale B, Kleidon J, Brinin A, Shand K, Waterhouse P, Dale J (2018) Gene editing the phytoene desaturase alleles of Cavendish banana using CRISPR/Cas9. Transgenic Res 27:451–460

Ortiz R (2013) Conventional banana and plantain breeding. Acta Hortic 986:177–194

Pegg KG, Langdon PW (1987) Fusarium wilt (Panama disease): a review. In: Persley GJ, De Langhe EA (eds) Banana and plantain breeding strategies. INIBAP, Cairns, pp 119–123

Pegg KG, Moore NY, Sorenson S (1993) Fusarium wilt in the Asian Pacific region. *Australian Journal of Agricultural Research* 47: 637–650

Ploetz RC (1994) Fusarium wilt and IMTP phase II. In: Jones DR (ed) The improvement and testing of *Musa*: a global partnership. INIBAP, FHIA, Montpellier, pp 57–59

Ploetz RC (2006) Panama disease: an old nemesis rears its ugly head. Part 2. The Cavendish era and beyond. Plant Health Prog. Online publication. https://doi.org/10.1094/PHP-2006-0308-01-RV

Ploetz RC (2015a) Fusarium wilt of Banana. Phytopathology 105:1512–1521

Ploetz RC (2015b) Management of Fusarium wilt of banana: a review with special reference to tropical race 4. Crop Prot 73:7–15

Ploetz RC, Correll JC (1998) Vegetative compatibility among races of *Fusarium oxysporum* f. sp. *cubense*. Plant Dis 72:325–328

Ploetz RC, Pegg KG (2000) Fusarium wilt. In: Jones DR (ed) Diseases of banana, Abaca and Enset. CABI Publishing, New York, pp 143–159

Ribeiro LR, Amorim EP, Cordeiro ZJM, De Oliveira E, Silva S, Dita MA (2011) Discrimination of banana genotypes for Fusarium wilt resistance in the greenhouse. Acta Hortic 897:381–386

Rowe P (1987) Banana breeding in Honduras. In: Persley GJ, De Langhe EA (eds) Banana and plantain breeding strategies. INIBAP, Cairns, pp 74–77

Smith MK, Hamill SD, Langdon PW, Giles JE, Doogan VJ, Pegg KG (2006) Towards the development of a Cavendish banana resistant to race 4 of Fusarium wilt: gamma irradiation of micropropagated Dwarf Parfitt (*Musa* spp., AAA group, Cavendish subgroup). Aust J Exp Agric 46:107–113

Smith LJ, Smith MK, Tree D, O'Keefe D, Galea VJ (2008) Development of a small-plant bioassay to assess banana grown from tissue culture for consistent infection by *Fusarium oxysporum* f. sp. *cubense*. Australas Plant Pathol 37:171–179

Stover RH (1962) Fusarium wilt (Panama disease) of bananas and other Musa species. Commonwealth Mycological Institute, Kew, Surrey, UK, 117 pp.

Tang CY, Hwang SC (1994) Musa mutation breeding in Taiwan. In: Jones DR (ed) Proceedings of the first global conference of the International Musa Testing Program. INIBAP, FHIA, Montpellier, pp 219–227

Thangavelu R, Mostert M, Gopi M, Ganga Devi P, Padmanaban B, Molina AB, Viljoen A (2019) First detection of *Fusarium oxysporum* f. sp. *cubense* tropical race 4 (TR4) on Cavendish banana in India. Eur J Plant Pathol 154:777–786

Viljoen A, Mahuku G, Massawe C, Ssali RT, Kimunye J, Mostert G, Ndayihanzamaso P, Coyne DL (2017) Banana pests and diseases: field guide for disease diagnostics and data collection. International Institute of Tropical Agriculture (IITA), Ibadan

Viljoen A, Ndayihanzamaso P, Mostert G (2018) Greenhouse inoculation of banana plantlets for Fusarium wilt resistance. Online publication: https://breedingbetterbananas.org/index.php/documents/greenhouse-inoculation-of-banana-plantlets-for-fusarium-wilt-resistance/. 29 Feb 2020

Viljoen A, Mostert D, Chiconela T, Beukes I, Fraser C, Dwyer J, Murray H, Amisse J, Matabuana E, Tazan G, Amugoli OM, Mondjana A, Vaz A, Pretorius A, Bothma S, Rose L, Beed F, Dusunceli F, Chao C-P, Molina AB (2020) Occurrence and spread of the banana fungus *Fusarium oxysporum* f. sp. *cubense* TR4 in Mozambique. S Afr J Sci 116. https://doi.org/10.17159/sajs.2020/8608

Warman NM, Aitken EAB (2018) The movement of *Fusarium oxysporum* f. sp. *cubense* (subtropical race 4) in susceptible cultivars of banana. Front Plant Sci 9:1748. https://doi.org/10.3389/fpls.2018.01748

Wu YL, Yi GJ, Peng XX (2010) Rapid screening of Musa species for resistance to Fusarium wilt in an in vitro bioassay. Eur J Plant Pathol 128:409–415

Chapter 6
Lab-Based Screening Using Hydroponic System for the Rapid Detection of Fusarium Wilt TR4 Tolerance/Resistance of Banana

Norazlina Noordin, Affrida Abu Hassan, Anis Nadia Mohd Faisol Mahadevan, Zaiton Ahmad, and Sakinah Ariffin

Abstract Field-based screening and evaluation of banana plant tolerance or resistance to *Fusarium oxysporum* f. sp. *cubense* (Foc) Tropical Race 4 (TR4) or also known as Fusarium wilt TR4 is ideal though not always feasible. Alternatively, screening of banana plantlets at lab-stage seems to be an effective method for early detection of Foc TR4 tolerance. We present a simple hydroponic system, that allows plant to grow in a water-based condition. The system has two layers, the upper layer is a tray that has holes for plantlets to be placed where the root system is supported using an inert medium such as rock-wool. The lower layer is a perforated container filled with a water-based nutrient solution. For this lab-based screening, *ex vitro* gamma irradiated banana cv. Berangan (AAA) rooted plantlets with a pseudostem height of 10–15 cm were inoculated by soaking in a Foc TR4 conidial suspension (10^6 spores/ml) for 2 h under room temperature. The Foc TR4 inoculated rooted plantlets were screened using the hydroponic system and disease symptoms were scored. In this chapter, protocols on acclimatization of *ex vitro* irradiated rooted plantlets, inoculation with a Foc TR4 conidial suspension, lab- screening using hydroponic system, observation for early detection of disease symptoms and scoring of disease severity are presented.

Keywords Banana · Gamma irradiation · *Fusarium oxysporum* f. sp. *cubense* · Lab-based screening · Hydroponic system · Disease severity scale · Resistance scoring

N. Noordin (✉) · A. A. Hassan · A. N. M. F. Mahadevan · Z. Ahmad · S. Ariffin
Agrotechnology and Biosciences Division, Malaysian Nuclear Agency, Ministry of Science, Technology and Innovation Malaysia (MOSTI), Kajang, Selangor, Malaysia
e-mail: azlina@nuclearmalaysia.gov.my

© The Author(s) 2022
J. Jankowicz-Cieslak, I. L. Ingelbrecht (eds.), *Efficient Screening Techniques to Identify Mutants with TR4 Resistance in Banana*,
https://doi.org/10.1007/978-3-662-64915-2_6

1   Introduction

Banana is the second most commonly grown fruit crop in Malaysia. About 50% of the banana growing area is cultivated with Pisang Berangan and the Cavendish type. However, in the recent years, the overall banana production and cultivation areas have decreased due to the increasing threat of Fusarium wilt and Moko diseases. This alarming issue has led to increasing prices and limited availabilities of banana fruits for local consumption and export (MOA report 2017).

Fusarium wilt of bananas, caused by *Fusarium oxysporum* f. sp. *cubense* (Foc), is one of the most devastating diseases of banana in most parts of the world including Malaysia. This severe disease almost crippled the world banana plantation, production and export trade of 'Gros Michel' circa 1940s and 1950s due to high susceptibility of this cultivar to Fusarium Race 1 (Pegg and Langdon 1987). Introduction of the resistant Cavendish group of cultivars provided an effective and economical solution but the emergence of Foc TR4 and its dissemination in the tropics and sub-tropics poses an immediate threat (Asif and Mak 2001). In Malaysia, we have experienced a total wipe out of a commercial scale Cavendish plantation in the southern part of Malaysia during the 1990s due to the destructive Foc TR4 disease.

Various control measures such as injection of chemicals, soil treatments including fumigation and incorporating soil amendments may reduce the severity of the disease, but none of them is commercially applicable (Hwang and Ko 1987). It was also reported that fungicides, fumigants, flood fallowing, crop rotation, and organic amendments have rarely provided long-term control in any production area (Pegg et al. 1996). Among three races of Foc namely Race 1, Race 2 and Tropical Race 4 which attack bananas, Tropical Race 4 is the most pathogenic and affects many banana cultivars including Cavendish. The pathogen, Foc, can be disseminated through suckers, soil, water, and by farming practices when farmers use contaminated tools. Chemical control, such as soil fumigation, is a promising measure but is hazardous to the environment. The pathogen persists in the contaminated soil by producing chlamydospores even in the absence of the host bananas or sometimes by infection of roots of some weeds (Pegg et al. 1996). As a result, once the field is invaded by Foc, the field cannot be used for banana production for up to 30 years (Asif and Mak 2001). It is therefore of high importance to develop and select new banana varieties that are resistant to Fusarium wilt to overcome this problem (Rusli 2011). One of the techniques to produce new varieties is through *in vitro* mutagenesis with the application of gamma irradiation (Mak et al. 1995). Gamma irradiation is a reliable and popular physical mutagen to increase genetic diversity and induce mutations and has led to the establishment of new plant varieties (Norazlina et al. 2014).

Field evaluation is the most reliable method for screening disease-resistant lines, but both manpower and space requirements are the limitations that add to the cost of screening (Pegg et al. 1996). It is also essential to maintain strict quarantine measures to avoid pathogen spread and cross-contamination. In addition, plants tend not to show disease symptoms until after 4–5 months (Morpurgo et al. 1994). The uneven distribution of the pathogen in the field can lead to 'disease escape' while many variables that can affect infection and symptom expression cannot be altered nor controlled (Asif and Mak 2001). Alternatively, methods that are simpler, cost effective, low maintenance with space requirements that can show early detection against Fusarium wilt are more attractive and have been developed by many laboratories.

Earlier studies reported the need for improved methods like pre-screening or early detection of tolerance against Fusarium wilt using *ex vitro* or *in vitro* rooted plantlets. This early detection is not only for screening for tolerance/resistance but also for comparative virulence and pathogenicity studies (Buddenhagen 1987; Pegg and Langdon 1987). This early detection can be carried out at lab-based or nursey-based stages prior to field-screening.

An earlier study reported roots inoculated by dipping or soaking with a fungal spore suspension before being transferred to an infested field (Mak et al. 2001; Vakili 1965). Susceptible banana plantlets showed external symptoms of leaf yellowing within 2 weeks and wilted within 4 weeks of inoculation (Vakili 1965). Earlier, a double-cup sand-culture containment method had been developed for testing pathogen virulence (Liew 1996). Modification was done and it was replaced by a 'double compartment' apparatus, which contains two plastic trays, one fitting inside the other. This double-tray technique has the capacity for pathogen containment to eliminate cross-contamination (Mak et al. 2001). This technique can be further modified to investigate the effects of various inoculum concentrations and environment variables on infection and disease expression. A bigger size tray could screen a greater number of plants. However, this system must be done in a nursery that still requires a large space with special containment, manpower for the preparation of the system and the cost for the set-up and modification is also less effective.

In this study, a lab-based screening using hydroponic system has been developed. This system is simpler, soilless, and easy to set up, cost effective, portable, requires less space and maintenance. The system can be modified to suit the requirements of the work in the laboratory. In this chapter, protocols on acclimatization of *ex vitro* irradiated rooted plantlets, inoculation with Foc TR4 conidial suspension, lab- based screening using hydroponic system and observation for early detection of disease symptoms and scoring of disease severity are presented. This work also describes the methodology, reliability of this lab-based screening by considering several factors, including the concentration and duration of Foc TR4 conidial suspension inoculation, type of host plants, and the ability to show differential disease symptoms similar to nursery-based and field evaluations.

2   Materials

2.1   Acclimatization of Banana Plantlets

1. Irradiated banana cv. Berangan (AAA) *in vitro* rooted plantlets with the pseudostem height of 5–7 cm (*see* Notes 1 and 2).
2. Ventilated culture containers (*see* Note 5).
3. Perlite (*see* Note 6).
4. Forceps.

2.2   Fusarium oxysporum f. sp. cubense (Foc) Tropical Race 4 (TR4) Cultures

1. Potato dextrose agar (PDA) medium.
2. 1% yeast extract.
3. Foc TR4 stock culture (*see* Note 7).

2.3   Foc Conidial Suspension

1. Sterile distilled water.
2. Sterile glass rod.
3. Filter funnel.
4. Cotton wools.
5. 500 ml glass beaker.
6. Haemocytometer.

2.4   Soaking in Foc TR4 Conidial Suspension

1. Foc TR4 suspension (10^6 spores/ml).
2. *Ex vitro* irradiated cv. Berangan rooted plantlets (10–15 cm pseudostem height) (*see* Note 1).
3. Soaking containers.
4. Biosafety Cabinet Level 2 (*see* Note 8).

2.5   Screening in Hydroponic System

1. Hydroponic system (*see* Note 10).
2. Distilled water.

3. Foc inoculated plantlets.
4. Nutrient rich solution (*see* Note 11).

3   Methods

3.1   Preparation of Acclimatized Ex Vitro Plantlets

1. Select healthy, contamination-free *in vitro* rooted plantlets for acclimatization phase (*see* Fig. 6.1).
2. Gently remove *in vitro* rooted plantlets with the pseudostem height of 5–7 cm from rooting media (*see* Fig. 6.1, Note 3).
3. Wash under running tap water to remove any traces of adhering agar/media.

Fig. 6.1 Berangan *in vitro* rooted plantlets selected for acclimatization. (**a**) Rooted Berangan cultures in rooting media. (**b**) *In vitro* rooted plantlets with the pseudostem height of 5–7 cm. (**c**) Washing and removal of adhering agar/media from *in vitro* rooted plantlets. (**d**) Transfer of plant material into ventilated culture containers containing perlite

Fig. 6.2 Acclimatization of Berangan *ex vitro* rooted plantlets. (**a**) Ventilated culture containers placed in hardening room. (**b**) 3-weeks old acclimatized Berangan rooted plantlets. (**c**) Berangan rooted plantlets with pseudostem height of 10–15 cm ready to be advanced to lab-based screening protocol using hydroponic system

4. Transfer plantlets using forceps into ventilated culture containers containing perlite. Each ventilated container can be filled with 12–15 plantlets (*see* Fig. 6.1).
5. Place the ventilated culture containers in hardening room at room temperature for 3 weeks (*see* Fig. 6.2).
6. After 3 weeks, *ex vitro* rooted plantlets with pseudostem height of 10–15 cm are ready to be advanced to lab-based screening protocol using hydroponic system (*see* Fig. 6.2).

3.2 Subculturing Foc TR4

1. Subculture Foc TR4 onto fresh PDA medium by taking one agar slab (0.5 × 0.5 cm) containing the fungal mycelium (*see* Fig. 6.3).
2. Transfer agar slab onto PDA medium supplemented with 1% yeast extract for further multiplication.
3. Incubate cultures under continuous light at 23–25 °C in growth chamber.
4. Observe the growth of Foc mycelium. Growth usually will take about 5–7 days (*see* Fig. 6.3).

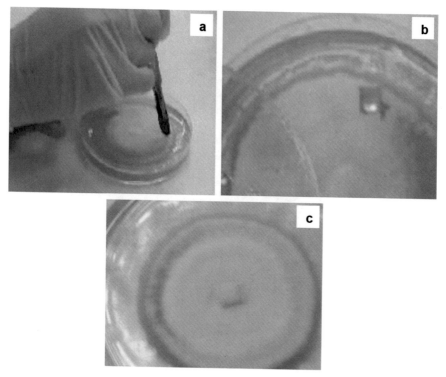

Fig. 6.3 Foc TR4 cultures. (**a**) Subculture of Foc TR4. (**b**) One agar slab (0.5 × 0.5 cm) cut. (**c**) Growth of Foc mycelium onto PDA medium

3.3   Preparation of Foc Conidial Suspension

1. Pour 10 ml of sterile distilled water onto PDA plates containing Foc mycelium and scrape all mycelium using a sterile glass rod.
2. Set up a filter funnel with plugging cotton wools on the mouth of a 500 ml glass beaker.
3. Pour the suspension through the funnel to separate hyphae from spores in the filtrate.
4. Using a haemocytometer count the number of spores/ml.
5. Dilute the suspension into desired concentrations (10^6 spores/ml) using sterile distilled water (*see* Fig. 6.4).

Fig. 6.4 Foc TR4 suspension (10^6 spores/ml) ready to be used for lab-based screening

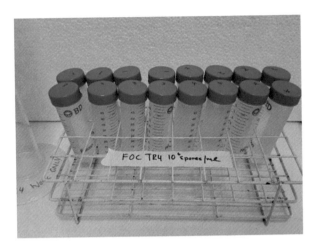

3.4 Soaking of Rooted Plantlets in Foc TR4 Conidia Suspension (10^6 Spore/ml)

1. Remove *ex vitro* rooted plantlets with pseudostem height of 10–15 cm after 3 weeks acclimatization in perlite (*see* Fig. 6.5).
2. Wash roots under running tap water to remove any traces of adhering perlite (*see* Fig. 6.5).
3. *Ex vitro* rooted plantlets with pseudostem height of 10–15 cm ready for pathogenicity test (*see* Note 13).
4. Inoculate root by soaking in the Foc TR4 conidia suspension (10^6 spore/ml) (*see* Fig. 6.6).
5. Conduct 3 treatments with different soaking periods for pathogenicity test (*see* Fig. 6.6):

   (a) T0: Control (non-inoculated with Foc TR4);
   (b) T1: 10^6 spore/ml *Fusarium* solution for 1 h;
   (c) T2: 10^6 spore/ml *Fusarium* solution for 2 h;
   (d) T3: 10^6 spore/ml *Fusarium* solution for 3 h.

3.5 Screening in Hydroponic System

1. Fill the hydroponic system with distilled water. In this type of hydroponic system, 15 l of water is used (*see* Fig. 6.7).
2. Gently transfer inoculated plantlets from all treatments into thumb-pots of the hydroponic system for screening of Foc TR4 disease symptoms (*see* Fig. 6.7).

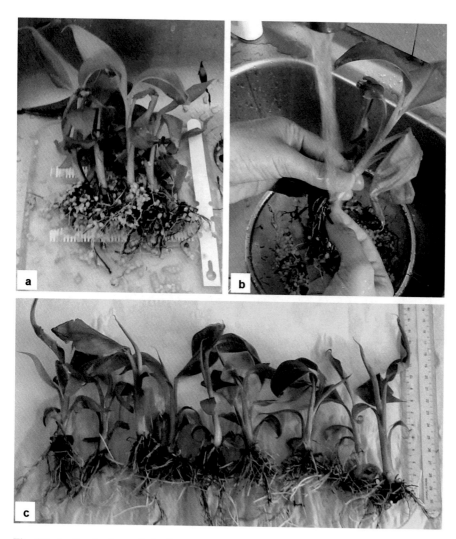

Fig. 6.5 Acclimatized rooted plantlets ready to be soaked in Foc TR4 conidial suspension. (**a**) *Ex vitro* rooted plantlets removed from perlite. (**b**) Washing roots to remove adhering perlite. (**c**) Rooted plantlets with pseudostem height of 10–15 cm ready for pathogenicity test

3. Screening for early detection against Foc TR4 is to be conducted in hardening room under natural conditions (*see* Fig. 6.8).
4. Symptom observations should be conducted weekly.

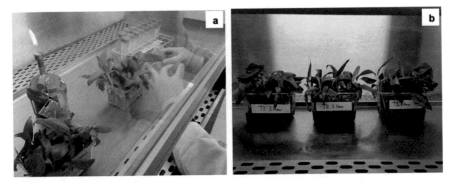

Fig. 6.6 Inoculation of roots with Foc TR4. (**a**) Soaking the irradiated rooted plantlets with Foc TR4 conidia suspension (106 spore/ml). (**b**) Soaking with different intervals (hours) under biosafety cabinet level 2

Fig. 6.7 Screening in hydroponic system. (**a**) Fill 15 l water in the lower compartment of the hydroponic system. (**b**) Transfer inoculated plantlets into thumb-pots of the hydroponic system

3.6 *Observation of Disease Symptoms*

1. Conduct weekly observations for disease development (28 days period).
2. Visual screening and observation of disease symptoms on weekly basis.
3. Rate disease severity on a scale of 0 to 6 (Wu et al. 2010) (*see* Table 6.1).
4. Select the workable and efficient treatment amongst the three different periods.
5. Proceed with the chosen treatment for further lab-based screening using a bigger population of inoculated plantlets (preferably 1000–2000) (*see* Note 14).

Fig. 6.8 Screen for early detection of resistance against Foc TR4

Table 6.1 Disease severity scale (Wu et al. 2010)

Disease severity index	Disease symptoms
0	Corm completely clean, no vascular discoloration
1	The smaller leaves at the base of pseudostem wilted, there was no discoloration of the pseudostem
2	≤1/2 the height of the pseudostem was discolored
3	>1/2 the height of the pseudostem was discolored and (or) there was discoloration of the leaf stalk
4	≤50% of the leaves wilted or yellowed
5	>50% of the leaves wilted or yellowed
6	Total discoloration of vascular tissue

3.7 Observation and Results from Lab-Based Screening Using Hydroponic System

1. After 28 days of inoculation with Foc TR4 (10^6 spore/ml) with different inoculation periods (1 h, 2 h, 3 h), conduct scoring of disease symptoms (*see* Fig. 6.9b).
2. Dissect plantlets from all treatments for scoring/rating (*see* Fig. 6.9c).

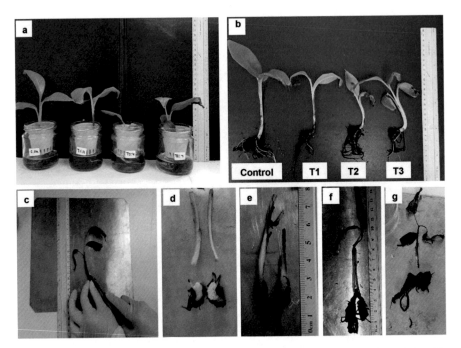

Fig. 6.9 Disease scoring (Wu et al. 2010). (**a**) Rooted plantlets after 28 days of inoculation. (**b**) Conduct scoring of disease symptoms for all treatments. (**c**) Plantlets cut-open for scoring. (**d**) Corm completely clean, no vascular discoloration (control). (**e**) ≤1/2 the height of the pseudostem was discolored (T3). (**f**) >1/2 the height of the pseudostem was discolored and (or) there was discoloration of the leaf stalk (T1). (**g**) Total discoloration of vascular tissue (T2)

3. For treatment 0 (control), the irradiated plantlets that were not inoculated with Foc conidial suspension showed no disease symptoms and these plantlets developed into healthy seedlings (*see* Fig. 6.9d).

4. Rate degrees of disease symptoms in T1 (1 h soaking period), T2 (2 h soaking period) and T3 (3 h soaking period) in Foc suspensions (Fig. 6.10).

5. Observe discoloration of pseudostem of the treated plantlets (*see* Fig. 6.9e–g).

6. Score plantlets from all treatments based on leaf symptoms responses (Brake et al. 1995) (*see* Fig. 6.11).

7. Banana plantlets showed external symptoms of leaf yellowing within 10–14 days.

8. The surviving plants showed disease scoring of 1–2.

   (a) Score 1: No streaking or yellowing of leaves. Plant appears healthy.
   (b) Score 2: Slight streaking and/or yellowing of lower leaves.

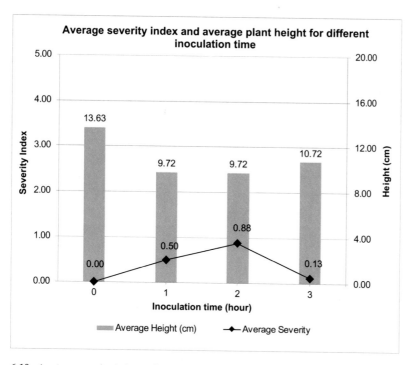

Fig. 6.10 Average severity index and average plant height for different inoculation time based on the disease scoring (Wu et al. 2010). Severity at the scale of 1 to 2 for T1 and T2. Average severity index was at 0.50 for treatment 1 and 0.88 for treatment 2. Scoring of 0 to 1 was observed in T3 treatment with average severity index of 0.13. Plant development (height) was decreased for both T1 and T2 treatment, with an average of 9.72 cm as compared to the non-inoculated plantlets; 13.63 cm

4 Notes

1. Mutagenesis of banana (cv. Berangan Intan) plantlets was performed by irradiating meristem tissues/shoot tips with acute gamma irradiation using BioBeam GM8000 (Germany) with a Caesium-137 source and chronic gamma irradiation using Gamma Greenhouse with a Caesium-137 source. The meristem pieces (about 1 cm × 2 mm) were aseptically excised from *in vitro* plantlets of Pisang Berangan. Each meristem was cut longitudinally into two pieces. A total of 50 meristem pieces for each dose were transferred into sterile moist Petri dishes and sealed with parafilm. The meristems in the Petri dishes were irradiated with acute gamma ray using gamma cell BioBeam GM8000 and chronic gamma irradiation using Gamma Greenhouse with a Caesium-137 source at 0, 10, 20, 30, 50, 70, 90, 120 Gy. Each growing irradiated shoot from all the optimal doses (chronic irradiation: 30, 50, 70 Gy and acute irradiation: 10, 20, 30 Gy) were separately sub-cultured to M_1V_4 generation (three subcultures at monthly interval) to minimize chimerism.

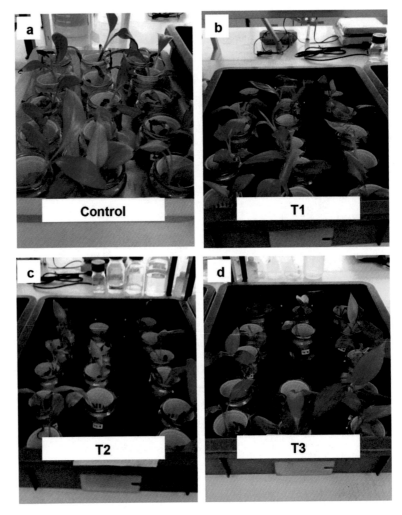

Fig. 6.11 Disease scoring observation on leaves symptomatic responses after inoculation in Foc suspension. (**a**) Non-inoculated plantlets showed no leaves yellowing. (**b**) T1 – 1 h soaking period. (**c**) T2 – 2 h soaking period. (**b**) T3 – 3 h soaking period

2. Irradiated plantlets (after M_1V_4) were continuously maintained in multiplication medium, modified MS Medium (Murashige and Skoog 1962) supplemented with 2.5 mg/l BAP and 0.1 mg/l NAA for further proliferation and plantlets multiplication.
3. For root induction, *in vitro* plantlets with the pseudostem height of 3–5 cm were cultured in MS medium supplemented with 1 mg/l IBA + 0.1% activated charcoal. Cultures were incubated at 24 °C with a 16 hours photoperiod (3500 lux).

4. All *in vitro* experiments should be conducted in a dedicated Plant Tissue Culture laboratory that should have a Laminar Air Flow cabinet and must adhere to proper decontamination procedures and disposal of waste.

5. The ventilated culture containers used during acclimatization of *ex vitro* rooted plantlets provide high humidity that is needed to help the plantlets to initiate the development of cuticles, stomata and root functions for better regeneration.

6. Perlite is used during acclimatization phase because it provides a soilless growth medium that can improve aeration, well-drained and eliminate any source of cross-contamination if soil-like growth medium is used.

7. The initial Foc TR4 cultures were kindly provided by the Plant Pathology Unit of Malaysian Agriculture Research and Development Institute (MARDI). The initial Foc TR4 cultures were sampled by MARDI from infected Berangan plants, the strain was tested by Koch Postulate test and were used to formulate the Foc TR4 hotspot for screening and evaluation against Foc TR4 disease. The Foc stock cultures were maintained by routine subculturing onto fresh PDA media and maintained under continuous light at 23–25 °C in a growth chamber.

8. All Foc TR4 related work and experiments i.e.; subculturing, preparation of conidial suspension and soaking/inoculating of roots in conidial suspension should be conducted with caution using suitable Personal Protective Equipment (PPE), dedicated labwares, dedicated biosafety cabinet level 2 in a dedicated Plant Pathology laboratory.

9. All labwares, glasswares, waste, disposable items related with Foc work must be separated from others and must always adhere to strict decontamination procedures and waste disposal.

10. There are many types of hydroponic systems for screening Foc disease symptoms. One can choose a hydroponic system that is the most suitable to work with. The principle is to ensure that the inoculated roots are totally immersed in the water.

11. The hydroponic nutrient rich solution contains Calcium Nitrate, Iron Chelate, Kalium Nitrate, Mono-Kalium Hydrogen Phosphate, Magnesium Sulphate, Manganase Sulphate, Boric Acid, Zinc Sulphate, Copper Sulphate and Natrium Molybdate.

12. Hydroponic system eliminates cross-contamination and allows concurrent investigations of Foc isolates against a range of irradiated rooted plantlets population.

13. Pathogenicity experiment can be conducted as a prerequisite for the establishment of a rapid and efficient Foc TR4 lab-based screening method for early detection of disease symptoms. This experiment can test the effectiveness of the Foc TR4 suspension (10^6 spores/ml) to cause disease symptoms to banana plantlets. It can also optimize the Foc TR4 soaking periods (minutes, hours) and the conidial suspension concentration.

14. Lab-based screening on bigger population of 1000–2000 irradiated rooted plantlets is recommended. The surviving plants that showed some degree of tolerance with severity index of 1–2 (Brake et al. 1995) will be screened under

nursery-based condition and later proceed to field-screening for selection of possible mutant lines.

15. From this study, the early screening method using hydroponic system provided baseline data on the development and optimization of an efficient and workable lab-based screening of Foc disease.

16. Soaking of roots of the irradiated plantlets in Foc TR4 (10^6 spore/ml) for 2 h interval was found to be effective.

17. It gave early expression of symptoms, as early as 14 days and produced consistent symptoms during the screening.

18. Soaking of roots of the irradiated plantlets in Foc TR4 (10^6 spore/ml) for 2 h and screening using hydroponic system for early detection of Foc TR4 tolerance/ resistance is effective to be used for screening of a bigger population (1000–2000 irradiated plantlets).

19. The plantlets that survived after inoculation and showed some degree of tolerance i.e., disease scoring of 1–2 (Brake et al. 1995) are advanced to nursery-based screening before being selected as putative mutants for field-based screening at a Foc TR4 hotspot.

Acknowledgments Authors wish to thank Mr. Shuhaimi Shamsudin, Mr Ayub Mohamad and Mr. Norhafiz Talib for their assistance and services for the acute and chronic gamma irradiation. Our heartfelt thanks and appreciation for Miss Nashimatul Adadiah Yahya, Miss Nurhayati Irwan and Mr. Mohamed Najli Mohamed Yasin for their dedication, endless support and assistance in the plant tissue culture laboratory. We would also like to thank the Plant Pathology Unit of Malaysian Agricultural Research and Development Institute (MARDI) for providing us the initial Foc TR4 mother-culture. Special thanks to the management of Agrotechnology and Biosciences Department and Malaysian Nuclear Agency for their continuous support of our R&D. Funding for this work was provided by the Food and Agriculture Organization of the United Nations and the International Atomic Energy Agency through their Joint FAO/IAEA Centre of Nuclear Techniques in Food and Agriculture. This work is part of IAEA Coordinated Research Project D22005.

References

Asif MJ, Mak C (2001) Characterization of Malaysian wild bananas based on anthocyanins. Biotropia 16:28–38

Brake VM, Pegg KG, Irwin JAG, Chaseling J (1995) The influence of temperature, inoculum level and race of Fusarium oxysporum f. sp. cubense on the disease reaction of the banana cv. 'Cavendish'. Aust J Agric Res 46:673–685. https://doi.org/10.1071/AR9950673

Buddenhagen IW (1987) Disease susceptibility and genetics in relation to breeding of banana and plantains. In: Persley GJ, De Langhe EA (eds) Banana and plantain breeding strategies, ACIAR proceedings no. 21. ACIAR, Canberra, pp 95–109

Hwang S, Ko WH (1987) Somaclonal variation of bananas and screening for resistance to Fusarium wilt. In: Persley GJ, De Langhe EA (eds) Banana and plantain breeding strategies, ACIAR proceedings no. 21. ACIAR, Canberra, pp 151–156

Liew KW (1996) Screening for disease resistance in banana plantlets against Fusarium wilt. Paper presented at the Regional training course on molecular approaches, mutation and other bio-technologies for the improvement of vegetatively propagated plants (FAO-UKM), 28 Oct– 8 Nov 1996, Bangi, Selangor, Malaysia

Mak C et al (1995) Mutation induction by gamma irradiation in a triploid banana Pisang Berangan, Malaysian. J Sci 16A:77–81

Mak C, Mohamed AA, Liew KW, Ho YW (2001) Early screening technique for Fusarium wilt resistance in banana micropropagated plants. Proceeding of a meeting held in Leuven, Belgium, 24–28 September 2001, pp 219–227

MOA et al (2017) 2017 report on statistic of Malaysian fruit crops. Ministry of Agriculture, Malaysia, Putrajaya

Morpurgo R et al (1994) Selection parameters for resistance to *Fusarium oxysporum* f. sp. *cubense* race 1 and race 4 on diploid banana (*Musa acuminata* Colla). Euphytica 75:121–129

Murashige T, Skoog F (1962) A revised medium for rapid growth and bioassay with tobacco tissue culture. Physiol Plant 15:473–497

Norazlina N, Rusli I, Nur Hidayah MS, Salmah M, Sobri H (2014) Effects of acute and chronic gamma irradiation on *in vitro* growth of *Stevia rebaudiana* bertoni. Jurnal Sains Nuklear Malaysia 2(2):17–25

Pegg KG, Langdon PW (1987) Fusarium wilt (Panama disease) – a review. In: Persley GJ, De Langhe EA (eds) Banana and plantain breeding strategies, ACIAR proceedings no. 21. ACIAR, Canberra, pp 119–123

Pegg KG et al (1996) Fusarium wilt of banana in Australia: a review. Aust J Agric Res 47:637–650

Rusli I (2011) FNCA report on mutation breeding achievement on disease resistance in Banana. Forum for Nuclear Cooperation in Asia, Tokyo, pp 9–13

Vakili NG (1965) Fusarium wilt resistance in seedlings and mature plants of *Musa* species. Phytopathology 55:135–140

Wu YL, Peng XX, Yi GJ (2010) Rapid screening of Musa species for resistance to Fusarium wilt in an in vitro bioassay. Eur J Plant Pathol 128(3):409–415

Chapter 7
Field Screening of Gamma-Irradiated Cavendish Bananas

Gaudencia A. Lantican

Abstract In our search for Cavendish bananas to withstand *Fusarium oxysporum* f. sp. *cubense* (Foc TR4) and other diseases, field screening of tissue-cultured Grand Nain banana seedlings derived from gamma-irradiated shoot tips was explored. Six months after irradiation and multiplication in the laboratory, the plantlets (M_1V_6) were individually grown in seedling bags under screen house conditions for 8 weeks, side-by-side with non-irradiated plantlets of the same clone. Once acclimatized, the banana plants were grown in an area confirmed positive of Foc TR4 (based on previous farm records stating that more than 50% of the plant population succumbed to the disease). Seedlings from each treatment (dose of radiation) were divided into four replicates, regardless of the number of plants. Each plant was given a unique identification code for traceability during disease monitoring, bunch and fruit quality evaluation.

Incidences of Foc TR4, Moko disease (*Ralstonia solanacearum*) and virus diseases were monitored weekly. Plants found positive of any disease were eradicated immediately. The plant population for the succeeding generation was managed by removing the unwanted suckers, 12 weeks from planting using a spade gouge and keeping only one sucker per plant for the next generation. Agronomic characters of each plant were taken at the flowering stage. These included age to flower, height, pseudostem circumference, number of leaves and height of the sucker. The bunch was harvested 12 weeks from flowering. The number of hands in a bunch, the number of fingers and weight of a hand were recorded. The same agronomic characters of the plant were taken for the succeeding generations.

Plants left standing in the field without any disease symptoms 3 years after planting were considered as putative mutants and were selected as candidate lines for multiplication and second-generation field screening. Only healthy suckers (free from viruses) were further multiplied *via* tissue culture technique to reach M_1V_6. Clean suckers from each line free of soil debris or dirt were sent to the laboratory for multiplication. At least 1000 plantlets were produced from each line for the

G. A. Lantican (✉)
Dole Philippines, Inc., Davao City, Philippines
e-mail: Gaudencia.Lantican@doleintl.com

© The Author(s) 2022
J. Jankowicz-Cieslak, I. L. Ingelbrecht (eds.), *Efficient Screening Techniques to Identify Mutants with TR4 Resistance in Banana*,
https://doi.org/10.1007/978-3-662-64915-2_7

second-generation field screening. These were grown in two locations – with and without records of Foc TR4. Field monitoring activities including plant population management, disease incidence assessment and fruit quality evaluation were carried out following the same protocols used in the establishment of the first-generation plants. Lines with population showing $\leq 10\%$ Foc TR4 after the first harvest, with good vigor, fruit quality and productivity were considered as candidates for further multiplication, farmers distribution and field planting under semi-commercial scale.

Keywords Gamma irradiation · Shoot tips · Field screening · Foc TR4 · Cavendish

1 Introduction

The choice of planting Cavendish bananas in the Philippines particularly Grand Nain resides in its high-yielding capacity, fruit quality and export market demands. After decades of cultivation, the fungal disease, *Fusarium oxysporum* f. sp. *cubense* (Foc TR4) debilitated the plant population, and its productivity per unit area. In some locations, the disease affects more than 50% of the population in a short period of time.

To date, there is no single effective treatment against the disease. The variety GCTCV 218 introduced from Taiwan Banana Research Institute (TBRI) through the Bureau of Plant Industry has helped to improve some devastated farms. Yet it has been found that this variety shows susceptibility to Foc TR4 at varying degrees, depending on the location. The availability of a resistant clone therefore, with qualities at par or better than Grand Nain is a welcome development to help manage, if not totally address the complexities of banana pest and disease problems. This will help growers and farmers especially in regions that are vulnerable and are positive of the disease.

Bhagwat and Duncan (1998) reported banana mutation breeding for tolerance to Foc TR4 using gamma irradiation. In 2006, Damasco et al. published a paper "Banana Bunchy Top Virus (BBTV) Resistance in Cultivar Lakatan Developed *via* Gamma Irradiation of Shoot Tips". This prompted Dole Research to seriously engage in experiments using Grand Nain clone of Cavendish. We have coordinated with the Philippine Nuclear Research Institute (PNRI) equipped with facilities to cover the gamma irradiation treatments. The protocol of Novak et al. (1990) served as an important guide in this project. Radiation treatment is at M_1V_1 stage of the shoot tips while inside the culture bottles and further multiplied to reach M_1V_6. Some aspects of Novak et al.'s (1990) protocol have been modified to better understand the characteristics of the plants in the field in addition to its susceptibility to diseases.

The field screening project started in year 2013 in Malandag, Malungon, Sarangani Province, and after 4 years, it was extended to two other regions namely, Maragusan and, Compostela Valley. Both locations are situated in the islands of Mindanao and Philippines.

This protocol describes the field screening tests of plants exposed to gamma irradiation treatments, their multiplication and further field screening tests of those that escaped from Foc TR4 and other diseases (putative mutants) after completing at least three harvest cycles. Disease incidence, plant vigor, growth and development patterns, productivity, and fruit quality were recorded. These parameters describe the tolerance of a putative mutant to a certain disease and ensure that fruit quality and productivity are within farmers' and consumers' level of acceptability.

2 Materials

2.1 Planting Material

1. Hardened banana seedlings (Fig. 7.1 and *see* Note 1).

2.2 Land Preparation and Planting

1. Farm record showing the historical Foc TR4 cases of the area, $\geq 50\%$ of the population.

Fig. 7.1 Establishment of healthy seedlings in the screen house for 8 weeks (**a**) and a healthy seedling during field planting (**b**)

2. Heavy equipment for land ripping and harrowing.
3. Soil penetrometer.
4. Transect line.
5. Wooden stakes.
6. Digging bags.
7. Shovels.
8. Fertilizer.
9. Calibrated scoops for fertilizer application.
10. Cutting knife.
11. Prepared plant labels (with treatment number, plant number).

2.3   Disease Survey and Eradication

1. Template pictures of disease symptoms.
2. Eradication solution: systemic herbicide (20%) in 1% ammonium sulfate solution.
3. Eradication injection equipment.

2.4   Bunch Care

1. Fruit bags.
2. Markers.
3. Data sheets.
4. Pens.
5. Clip boards.

2.5   Harvest and Transport

1. Ladder.
2. Harvesting knife.
3. Weighing scale.
4. Table.
5. Measuring tape.
6. Data sheets.
7. Pens.
8. Clip boards.
9. Paper boxes or wooden crates.

2.6 Fruit Quality Assessment

1. Digital weighing scale.
2. Oven.
3. Refractometer.
4. Ripening chamber.
5. Reefer container.
6. Color meter.

2.7 Selection and Transport of Putative Mutant Suckers from Field to Laboratory

1. Gouging bars.
2. Knife.
3. Bolo.
4. Labels and tags.
5. Pens.
6. Fine mesh to cover the suckers during transport.
7. Plastic sacks to contain the suckers.
8. Fine mesh to cover the suckers.
9. Refrigerated container or air-conditioned vehicle to transport the suckers from the field to the laboratory.

3 Methods

3.1 Selection of Land for Planting

1. Select an area which is confirmed to have Foc TR4 (*see* Note 2).

3.2 Land Preparation and Planting

1. Deep plough and rip at 50–60 cm depth to reach soil compaction to level < 2 Kg/cm^2 (*see* Note 3).
2. Plant in a single line pattern at population of 2222 per hectare at 3-m distance between lines and at 1.5-m distance between plants, orientated East to West (*see* Note 4).
3. Prepare planting holes of the size 20-cm wide and 30-cm depth (*see* Note 5).

4. Depending on the number of treatments (radiation doses) and the number of seedlings produced, the layout planting design is obtained by dividing the seedlings from each treatment to cover at least four replicates in the field.
5. Give a permanent and unique identification code to each plant after planting.
6. Prepare a field map to indicate full details of plant identification and location.
7. Record the date and week of the year of planting.

3.3 Disease Monitoring

1. The entire population is surveyed at least once a week.
2. Individual plant is inspected if showing any disease symptoms.
3. Diseases to be recorded include Foc TR4, Moko, virus diseases (BBTV, BBrMV, CMV) and other abnormalities.
4. Plants with disease are eradicated within 24 h from planting.

3.4 Disease Eradication

1. Surveyed plants with disease will be further confirmed by the eradicator (*see* Note 6).
2. If confirmed positive of the disease, the plant will be injected with 50–100 ml of the eradication solution.
3. If sucker/s are present, injection will also be performed using ~50 ml eradication solution.
4. Gather and contain the debris of eradicated plants within the same planting spot.

3.5 Plant Care and Population Management

1. At 6 weeks from planting, dried or burnt leaves due to sigatoka or natural senescence are trimmed.
2. Leaf trimming is done once every 2 weeks.
3. After 12 weeks from planting, each plant is surveyed for presence and density of suckers.
4. Count and record the number of suckers in a plant.
5. Select one sucker nearest to the mother plant to serve as the second-generation plant.
6. Remove the other suckers (unwanted suckers) using a spade gouge.
7. Repeat removal of unwanted suckers once every month until the plant starts to flower.
8. At flowering stage, select the third-generation sucker, or until available.
9. The monthly schedule of sucker removal is followed for the succeeding generations.

3.6 Bunch Care

1. At flowering stage, the following parameters are collected:

    (a) Number of leaves
    (b) Plant height
    (c) Height of the selected sucker (next generation)
    (d) Pseudostem circumference (1-m from the base)

2. When all the hands in a bunch are exposed, the bunch is cleaned by removing the dried flowers, bracts and the male flower.
3. Tag the bunch to indicate date of flowering and to track harvest dates.
4. Place plastic inserts in between fingers (if necessary) to avoid fruit bruises.
5. Cover the cleaned bunch with a bag (paper, fabric or polyethylene) if, and when available.
6. Protect the fruit-bearing plant by installing bamboo poles (and/or its equivalent) as propping materials.
7. Harvest the bunch 12 weeks from flowering (*see* Note 7).

3.7 Harvest

1. Using a special harvest knife, harvest the bunch by cutting one hand at a time starting from the last hand (*see* Note 8).
2. The following information should be recorded:

    (a) Number of hands
    (b) Number of fingers in a hand
    (c) Weight of each hand

3. Use the second and last hand of each bunch for fruit quality reading.
4. Pack and label fruits before dispatching them to the laboratory in a refrigerated condition (13.5 °C).
5. Leave the pseudostem of the harvested plant intact with the youngest two leaves (*see* Note 9).

3.8 Fruit Quality Reading

1. For each bunch, divide the second and last hand specimens into two.
2. The first half will be read for pulp moisture content at color green.

3. The second half will be subjected to standard ripening procedure.
4. Fruit will be read for peel thickness, pulp moisture content, total soluble solids and/or brix at color yellow, 4 days after gassing.

3.9 Overall Plant Population and Farm Status

1. After harvest of the first generation, the remaining plants free from Foc TR4 symptoms and other diseases are inventoried (Table 7.1).
2. Follow above methods Sects. 3.3 to 3.8 for the succeeding generations until two to three harvest cycles are completed.
3. When population left in the area is ≤20% from the initial population, allow two or more suckers to grow from each plant.
4. Assess the number of individual plants that survived from each treatment and of the entire experimental lot.
5. Each of this survived plant from the gamma irradiated meriplants are considered putative mutant lines.
6. Archive data on plant growth, vigor and fruit quality of each putative mutant to decide if characteristics are qualified for further field evaluation (*see* Note 10).

3.10 Handling of Putative Mutant Plants from an Infected Field for Laboratory Multiplication and Second-Generation Field Screening

1. Once the putative mutant plants are selected, the available suckers are tagged.
2. Obtain tissue samples to test and verify for fungal or virus diseases (if any) *via* PCR in the laboratory (Company Standards).
3. Only suckers free from any disease will be dug and cleared from soil or dirt and sent to the laboratory.

Table 7.1 An illustration of survived plants in the field after 3 years of screening seedlings from shoot tips subjected to gamma irradiation

Doses of gamma radiation (Gy)	Number of seedlings planted	Number of putative mutant plants free of Foc TR4, moko and virus diseases		
		52 weeks after planting	104 weeks after planting	156 weeks after planting
0	96	0	–	–
8	168	4	1	1
10	278	6	2	0
15	113	2	2	2

Table 7.2 Seedlings multiplied from the putative mutant plants for second generation field screening

Putative mutant plant code	Number of seedlings produced	Planting locations	
		(1) Positive of Foc TR4	(2) Negative of Foc TR4
Control	1000	500	500
(a) 8 Gy plant #1[a]	1000	500	500
(b) 8 Gy plant #2[a]	1000	500	500
(c) 10 Gy plant #1[a]	1000	500	500
(d) 10 Gy plant #2[a]	1000	500	500
(e) 15 Gy plant #1[a]	1000	500	500

[a]Selected putative mutant plants showing desirable agronomic characteristics and fruit quality

4. Maintain a quarantine condition during transport of suckers by containing individually in a sack and enclosing with a fine net.
5. Transport suckers in a refrigerated van (if possible) or inside an air-conditioned vehicle.
6. Process the selected suckers immediately upon arrival in the laboratory.
7. Follow the standard tissue culture procedure to reach M_1V_6.
8. If untreated plants are not available in the experimental field, source the untreated plants of the same clone (Grand Nain) and include them in the batch of tissue culture intended for second generation field screening.
9. Produce at least 1000 plantlets from each line (*see* Note 11).
10. At M_1V_6 stage, submit sample plantlets from each line to a reputable laboratory, side-by-side with the untreated control (original clone) to identify presence or absence of mutations.
11. In the screen house, nurture the plantlets into healthy seedlings for 8 weeks, following the standard commercial practices.
12. Plant the 8-week-old seedlings in two locations – (1) confirmed positive of Foc TR4 where ≥50% of the population succumbed to the disease; and (2) confirmed negative of Foc TR4 (*see* Note 12, Table 7.2).
13. Gather the same data and information collected in the first-generation planting.
14. Select at least 30 bunches from each line when the population reaches 25, 50 and 75% harvests for quality readings in the laboratory following Sect. 3.8 above. Figure 7.2 illustrates plants in a population of putative mutant lines under second generation field screening. Both plants are clones of Grand Nain but received different doses of gamma irradiation treatment and survived during the first-generation planting.
15. Summarize cases of diseases, plant growth and productivity data after completing at least 85% of the first harvest (*see* Note 13, Table 7.3 and Fig. 7.3).
16. Decide which line to further multiply for farmer distribution and commercial-scale field planting after completion of at least three harvest cycles.
17. Apply for a patent any line deemed important.

Fig. 7.2 Sample fruit under second generation field screening of putative mutant lines originally from one Cavendish clone (Grand Nain) at different doses of gamma radiation – (**a**) 8 Gy and (**b**) 10 Gy

Table 7.3 Sample field screening data from two different locations showing incidence of diseases, after completing the first-generation harvest

Putative mutant lines	Location 1 (Positive of Foc TR4)			Location 2 (Negative of Foc TR4)		
	% Foc TR4	% Moko	%Virus	% Foc TR4	% Moko	%Virus
Control	29.0	0	0	0	0	0
(a)	13.0	0	0	0	0	0
(b)	4.0	0	8.0	0	0	0
(c)	8.0	0	0	0	3.0	3.0
(d)	11.0	0	1.0	0	0	2.5
(e)	20.0	0	4.0	0	0	0

18. Register to IAEA database (Mutant Variety Database) as product of induced mutation breeding *via* gamma irradiation.
19. Submit candidate Foc TR4 tolerant line/s to International Transit Centre (https://www.bioversityinternational.org/banana-genebank/) for preservation and as backup specimens for further research investigations.
20. Perform molecular characterization of identified mutant lines (*see* Note 14).
21. Direct field screening of banana seedlings from gamma irradiated shoot tips is time consuming, laborious and expensive. From the point of view of a banana grower, the measure of significance of an introduced variety covers both its tolerance to the disease of interest and the resulting agronomic character of the novel variety. Our project was launched in 2013 and is estimated to take five to 10 years to achieve successful results, with great possibility that mutant line/s

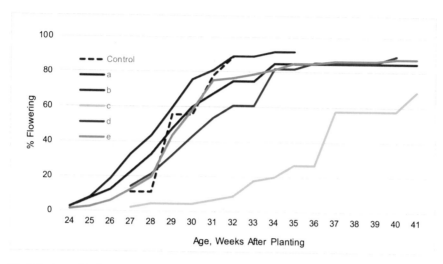

Fig. 7.3 A graphical presentation of the weekly cumulative flowering of putative mutant lines (*a*–*e*) in an area negative of Foc TR4, compared with the untreated control

Table 7.4 Program of activities and estimated time frame in the field screening of gamma irradiated Cavendish bananas for disease tolerance, plant vigor and fruit quality

	Activities	Time frame (weeks)
1	Multiplication in the laboratory of gamma irradiated shoot tips to produce plantlets (M_1V_6)	24
2	Growing of plantlets in the screen house to produce seedlings	8
3	First generation field screening of the produced seedlings	52–156
	(a) First harvest cycle (52 weeks)	
	(b) Second harvest cycle (104 weeks)	
	(c) Third harvest cycle (156 weeks)	
4	Selection of plants left standing in the field after 156 weeks as putative mutants	
	(a) Multiplication in the laboratory to produce plantlets, M_1V_6	24
	(b) Establishment of seedlings in the screen house	8
5	Second generation field screening (putative mutant lines)	52–156
	(a) Location 1 – Foc TR4 positive	
	(b) Location 2 – Foc TR4 negative	
6	Plant selections from the lines and further multiplication for commercial planting	
	(a) Multiplication in the laboratory to produce plantlets, M_1V_6	24
	(b) Establishment of seedlings in the screen house	8
	(c) Farmers distribution for semi-commercial field planting	52–156
Time Frame (weeks)		252–564

will be produced (*see* Table 7.4). In order to meet this target, the screening process for disease tolerance need to be planned and implemented correctly.

4 Notes

1. Seedlings from Grand Nain shoot tips were used as planting materials. They were subjected to different doses of gamma radiation, while the untreated control seedlings were grown in the screen house, provided with proper irrigation and nutrition for 8 weeks. Healthy seedlings are advantageous for quick recovery from stress during transport from the screen house and during field planting.
2. The area must be confirmed positive of Foc TR4 to challenge the response of the plants under natural, field conditions. Farm records must show that more than 50% of the host plants in the previous cropping succumbed to the disease in 1 year or less from planting.
3. Proper land preparation is required according to soil compaction data. It is believed that the process facilitates mixture of fungal inocula in the soil.
4. Planting pattern follows the commercial practice and population per hectare of the clone.
5. No other soil amendments are required, while fertilization follows the recommended protocol for Cavendish (Robinson and Sauco, 2010).
6. Both the disease surveyor and the disease eradicator must be knowledgeable of any disease symptoms or abnormalities of bananas. The confirmation by the eradicator will avoid mis-eradication of the plant and wrong record entry. Whenever a new disease symptom is observed, both the surveyor and eradicator report to the scientist/s for further confirmation.
7. If a plant with a hanging bunch is positive of any disease (Foc TR4, Moko or any virus disease), it will be eradicated similar to schedule and procedures described in Sect. 3.4 above.
8. If the plant is tall, a sturdy ladder will be required during harvest.
9. The natural decomposition of the pseudostem of the mother plant will facilitate growth of the succeeding generation.
10. It is the prerogative of the scientist to decide what other features, aside from disease tolerance are necessary to proceed with multiplication of a putative mutant line. In general, productivity (weight of the bunch and age from planting to harvest), moisture content and brix are important parameters to consider.
11. A 'line' is defined in this article as the resulting tissue culture plants obtained from a putative mutant.
12. Planting of putative mutant lines in an area confirmed free of Foc TR4 will help understand and establish the productivity potential of a new line.
13. Plant growth and vigor in an area free of Foc TR4 can be illustrated by the flowering pattern vis-à-vis control (Fig. 7.3). These can also be supported by productivity data such as bunch weights, number of hands in a bunch and fingers in a hand. Under commercial scale, a plant population of a line, variety or clone

with ≤10% Foc TR4 infection in the first cropping (from planting to completion of first harvest cycle) is a better alternative than the susceptible Grand Nain.

14. Samples of putative mutant lines have been submitted to Plant Breeding and Genetics Laboratory, Joint FAO/IAEA Centre on Nuclear Techniques in Food and Agriculture, Department of Nuclear Applications, International Atomic Energy Agency for molecular analysis together with the untreated control.

Acknowledgments   FAO/IAEA Co-ordinated Research Project on Efficient Screening Techniques to Identify Mutants with Disease Resistance for Coffee and Banana.
Plant Breeding and Genetics Laboratory, Joint FAO/IAEA Centre of Nuclear Techniques in Food and Agriculture, Department of Nuclear Sciences and Applications, IAEA – Dr. Ivan Ingelbrecht, Dr. Joanna Jankowicz-Cieslak, Dr. Stephan Nielen, Dr. Bradley Till.
The Philippine Nuclear Research Institute.
Dole Philippines, Inc. – Research, Production and Management Teams.

References

Bhagwat B, Duncan EJ (1998) Mutation breeding of Highgate (*Musa acuminata*, AAA) for tolerance to *Fusarium oxysporum* f. sp. *cubense* using gamma irradiation. Euphytica 101: 143–150

Damasco OP, Estrella JB, Caymo LS, Dizon TO, Rabara RC, Dela Cruz FS, Mendoza EMT (2006) Banana bunchy top virus (BBTV) resistance in cultivar 'Lakatan' developed via gamma irradiation of shoot tips. Philipp J Crop Sci 31(3):21–34

Novak FJ, Afza R, van Duren M, Omar MS (1990) Mutation induction by gamma irradiation of *in vitro* cultured shoot-tips of banana and plantain (Musa cvs). Trop Agric 67(1):21–28

Robinson JC, Sauco VG (2010) Bananas and plantains. In: Crop production science in horticulture, 2nd edn. CAB International, 311pp

Part III
Mutation Detection Using Genomics Tools

Chapter 8
Mutation Detection in Gamma-Irradiated Banana Using Low Coverage Copy Number Variation

Joanna Jankowicz-Cieslak, Ivan L. Ingelbrecht, and Bradley J. Till

Abstract Mutagenesis of *in vitro* propagated bananas is an efficient method to introduce novel alleles and broaden genetic diversity. The FAO/IAEA Plant Breeding and Genetics Laboratory previously established efficient methods for mutation induction of *in vitro* shoot tips in banana using physical and chemical mutagens as well as methods for the efficient discovery of ethyl methanesulphonate (EMS) induced single nucleotide mutations in targeted genes. Officially released mutant banana varieties have been created using gamma rays, a mutagen that can produce large genomic changes such as insertions and deletions (InDels). Such dosage mutations may be particularly important for generating observable phenotypes in polyploids such as banana. Here, we describe a Next Generation Sequencing (NGS) approach in Cavendish (AAA) bananas to identify large genomic InDels. The method is based on low coverage whole genome sequencing (LC-WGS) using an Illumina short-read sequencing platform. We provide details for sonication-mediated library preparation and the installation and use of freely available computer software to identify copy number variation in Cavendish banana. Alternative DNA library construction procedures and bioinformatics tools are briefly described. Example data is provided for the mutant variety Novaria and cv Grande Naine (AAA), but the methodology can be equally applied for triploid bananas with mixed genomes (A and B) and is useful for the characterization of putative Fusarium Wilt TR4 resistant mutant lines described elsewhere in this protocol book.

Keywords Physical mutagenesis · Banana · Polyploidy · TR4 · CNV · NGS

J. Jankowicz-Cieslak (✉) · I. L. Ingelbrecht
Plant Breeding and Genetics Laboratory, Joint FAO/IAEA Centre of Nuclear Techniques in Food and Agriculture, IAEA Laboratories Seibersdorf, International Atomic Energy Agency, Vienna International Centre, Vienna, Austria
e-mail: j.jankowicz@iaea.org

B. J. Till (✉)
Veterinary Genetics Laboratory, University of California, Davis, Davis, CA, USA

© The Author(s) 2022
J. Jankowicz-Cieslak, I. L. Ingelbrecht (eds.), *Efficient Screening Techniques to Identify Mutants with TR4 Resistance in Banana*,
https://doi.org/10.1007/978-3-662-64915-2_8

1 Introduction

New developments in the field of molecular biology enable fast and accurate identification of spontaneously occurring or induced changes of DNA sequence. This can allow a more precise use of induced mutations in crop improvement programmes. Mutagens produce various spectrum of changes. Chemical mutagens such as EMS predominantly induce point mutations, whereby physical mutagens such as gamma irradiation produce a broader spectrum of changes ranging from SNPs and small InDels to deletions greater than one million base pairs (Jankowicz-Cieslak and Till 2015; Till et al. 2018; Datta et al. 2018). While phenotypic consequences of large structural variants may be greater, datasets on the spectrum and density of such mutations are currently much smaller than that of EMS.

Mutation breeding may be especially useful in species with a narrow genetic base or those that are recalcitrant to traditional breeding methods such as obligate and facultative vegetatively propagated species. Additionally, mutagens that cause dominant or dosage-based phenotypes can increase the efficiency of generating novel traits in polyploids as the expression of phenotypes arising from recessive mutations requires the combination mutations from homologous sequences (Krasileva et al. 2017). Gamma irradiation has been used widely as a mutagenizing agent for breeding programmes for many crops. In poplar, treatment of pollen with gamma irradiation resulted in InDels varying between small fragments to whole chromosomes (Henry et al. 2015). This work further showed that large genomic InDels could be effectively recovered using low coverage whole genome sequencing (LC-WGS), making mutation discovery more cost-effective and data analyses more streamlined. Larger deletions range in size and may include loss of part of a chromosome (segmental aneuploidy) or loss of an entire chromosome (aneuploidy). Aneuploidy is better tolerated in polyploid plants and may be lethal for diploid plants and animals (Siegel and Amon 2012). These lead to changes in copy number of single genes to whole chromosomes, which have profound effects on phenotypes of the organism. Copy number variations especially affect haploinsufficient genes for which a single functional copy of a gene is not sufficient for normal function. Single copy mutations can potentially knock out the function of genes where only one functional copy is being maintained.

Inducing mutations in triploid banana provides an approach for generating novel variation that is heritable. The logistics of banana mutation breeding including tissue culture propagation, chimerism, polyploidy, heterozygosity, and field space required to find rare favourable mutations makes banana less tractable than seed propagated crops. However, these limitations can be overcome by tissue culture mutagenesis and genomic screening at earlier stages. Previously, we established a system for inducing and maintaining SNP mutations in clonally propagated banana plants. Treating shoot apical meristems of tissue cultured bananas with the chemical mutagen ethyl methanesulphonate (EMS) introduced a high density of GC-AT transitions mutations (Jankowicz-Cieslak et al. 2012). We further showed that mosaicism (chimerism) caused by accumulation of chemically induced mutations in different cells of the plant propagule could be rapidly removed *via* isolation of shoot apical meristems and subsequent longitudinal bisection. Further, induced mutations were maintained in mutant plants for more than six generations.

We sought to establish a similar system for inducing and maintaining insertions and deletions using physical irradiation. We aimed to develop an efficient pipeline for the generation and recovery of large copy number variations (CNVs) in gamma irradiated Cavendish banana cultivars, employing tissue culture, low coverage whole genome sequencing (LC-WGS) and chromosome dosage analysis (Fig. 8.1). We chose a chromosomal dosage analysis that was previously successful in detecting aneuploidy, insertions and deletions in Arabidopsis, rice and poplar (Tan et al. 2015, 2016). To establish a pipeline for banana, we first adapted sequencing and dosage analysis for the previously released mutant banana variety Novaria. Large genomic deletions of up to 3.8 Mbps were recovered. We next developed a newly mutagenized banana population and tested two different irradiation dosages to establish that new genetic variation can be induced and maintained *in vitro* (Datta et al. 2018). This work suggests that a large-scale mutagenesis pipeline can be

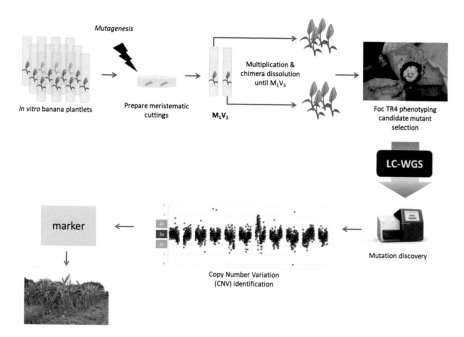

Fig. 8.1 A pipeline for the generation and recovery of large copy number variations (CNVs) in gamma irradiated Cavendish banana cultivars, employing tissue culture, low coverage whole genome sequencing (LC-WGS) and chromosome dosage analysis. An *in vitro* mutant population was generated, and a subset was evaluated using the method described in this chapter. This ensures that mutagenesis was successful and mitotically heritable DNA lesions were produced during gamma irradiation and subsequent propagation. Genome sequencing can also be applied to plants showing improved resistance to disease such as Foc TR4 in order to identify mutations causative for the observed phenotype(s). (This figure is modified from Jankowicz-Cieslak et al. 2021)

created for routine production of mutant populations suitable for glasshouse and field evaluations. The efficacy of this approach is being further tested for Foc TR4 resistance (Fig. 8.1). We provide here the methodology for low-coverage DNA sequencing and data analysis to identify large indels in mutant populations of triploid (AAA) banana.

2 Materials

2.1 Library Preparation and Sequencing

2.1.1 DNA Isolation and Quantification

1. DNA isolation kit (*e.g.* DNeasy Plant Mini Kit, Qiagen, Cat Nr: 69104).
2. Vortex mixer.
3. Microcentrifuge.
4. Micropipettes (1000 μl, 200 μl, 20 μl, 10 μl).
5. Microcentrifuge tubes (1.5 ml, 2.0 ml).
6. Metal beads (*e.g.* tungsten carbide beads, 3 mm, Qiagen, Cat Nr: 69997; *see* Note 1).
7. Distilled or deionized water (dH$_2$O).
8. RNase A (10 μg/ml).
9. Absolute ethanol.
10. 10× TE buffer (100 mM Tris-HCl, 10 mM EDTA, pH 8.0).
11. Equipment for horizontal gel electrophoresis (combs, casting trays, gel tank, power supply).
12. Agarose (for gel electrophoresis).
13. 0.5× TBE (Tris/Borate/EDTA) buffer (for gel electrophoresis).
14. Ethidium bromide or equivalent double stranded DNA dye (*see* Note 2).
15. Lambda DNA (*e.g.* Invitrogen, Cat Nr: 25250-010) for the preparation of concentration standards (*see* Note 3).
16. DNA gel loading dye (containing Orange G or bromophenol blue) (*see* Note 4).
17. Gel photography system (digital camera, light box).

2.1.2 Library Preparation and Sequencing

1. Covaris M220 Ultrasonicator.
2. microTUBE AFA Fiber Pre-Slit Snap-Cap (Cat Nr: 520077).
3. TruSeq Kit (*e.g.* TruSeq®Nano DNA Library Prep kit; *see* Note 5).
4. Fresh 70% EtOH.
5. Magnetic stand.
6. Qubit system (Thermo Fisher Scientific).

7. Qubit dsDNA HS Assay Kit (Cat Nr: Q32854).
8. PCR cycler.

2.2 DATA Analyses

1. Computer with minimum processor requirement (*see* Note 6).
2. Burrows-Wheeler Aligner (BWA) (http://bio-bwa.sourceforge.net/).
3. BBmap (https://sourceforge.net/projects/bbmap/).
4. Samtools (http://www.htslib.org/).
5. Bin-by-Sam-tool (*see* Note 7).
6. Python version 2.7 (pre-installed with the Ubuntu operating system).
7. Spreadsheet software (pre-installed with the Ubuntu operating system).

3 Methods

3.1 Library Preparation and Sequencing

3.1.1 DNA Isolation

1. Collect 100 mg fresh weight tissue per sample and freeze at −80° C (*see* Note 8).
2. Extract DNA using the DNeasy Plant Mini Kit (Qiagen, Cat Nr: 69106) or equivalent (*see* Note 9).

3.1.2 Assay DNA Quality and Quantity

1. Measure concentration using a Qubit fluorometer (*see* Note 10).
2. Prepare DNA concentration standards for gel electrophoresis (Huynh et al. 2017).
3. Dilute lambda DNA to a set of DNA concentrations covering the expected range of the genomic DNA samples being assayed (*see* Note 11).
4. Prepare a 1.5% agarose gel in 0.5× TBE buffer with 0.2 μg/ml ethidium bromide or alternative dye.
5. Add 3 μl of DNA sample plus 2 μl DNA loading dye (*see* Note 12).
6. Load samples and concentration standards on the gel.
7. Run gel at 5–6 V/cm for 30–60 min (*see* Note 13).

3.1.3 Library Preparation for Sequencing

1. Choose library preparation method, sequencing chemistry and read-length (*see* Note 14).

2. If using low-DNA input library preparation with 550 bp fragments for 2×300PE (paired-end) sequencing, add 500 ng of genomic DNA to a Covaris Snap-Cap microTUBE.

3. Shear the DNA in the M220 Covaris sonicator with the following settings: Peak Incident Power (W); 50, Duty Factor (%) 10; Cycles per Burst (cpb) 200; Treatment Time (sec) 50 (*see* Note 14).

4. Proceed to library preparation using a TruSeq®Nano DNA Library Prep kit or equivalent, following manufacturers protocol. Use unique barcodes/indices for each sample.

5. Quantify libraries using Qubit and determine molarity using the following equation where 660 is the molecular weight of a DNA base pair and median size is the average size of fragments in base pairs (*see* Note 15): ng/ul * 1 mol/(660)g * MEDIAN SIZE* 1 g/10ee9ng*10ee9nmol/mol*1ul/10ee-6 l = nmol/l

6. Determine sample pooling based on sequencing throughput and genome size (*see* Note 16).

7. Adjust each sample to 4 nM and pool samples together according to Note 16.

8. Sequence pooled library (*see* Note 17)

3.1.4 Data Analysis

1. Obtain Illumina raw sequence reads (*see* Note 18 and Fig. 8.2).
2. Set up your computer for analysis (*see* Note 6).

Fig. 8.2 Diagram of bioinformatics steps to recover candidate mutations from gamma irradiated bananas

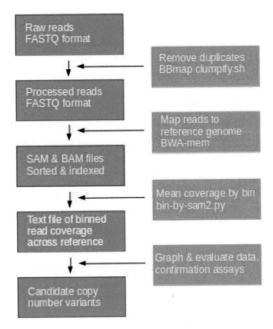

3. Install BWA following installation instructions on SourceForge. Alternatively, if running Ubuntu, the software is available from the repository and can be loaded by opening a terminal window and typing: sudo apt install bwa.

4. Install samtools following instructions on htslib.org. Alternatively, if using Ubuntu, type the command sudo apt install samtools into the terminal window.

5. Quality filter the sequence reads and trim to remove adapter sequences.

6. Identify samples according to index and de-multiplex (*see* Note 19).

7. Prepare the fastq data for mapping. In the example data provided (https://www.ncbi.nlm.nih.gov/bioproject/PRJNA627139), the sequencing library was prepared using a kit that includes a post adapter-ligation amplification step. This can produce duplicated reads that affect data analysis. Duplicate reads can be removed using a tool in the BBmap suite. Install BBmap following the instructions found at https://jgi.doe.gov/data-and-tools/bbtools/bb-tools-user-guide/installation-guide/. Place the processed fastq.gz into the BBmap directory. Open a terminal window (found in the applications folder in Ubuntu 18.04, enter (cd) the BBmap directory and then execute the reformat software. The command to execute this from the BBmap directory is

./clumpify.sh in = Sample.R1.fq.gz in2 = Sample.R2.fq.gz out = Sample.R1. dedup.fastq.gz out2 = Sample.R2.dedup.fastq.gz dedupe;
Where Sample.R1 and Sample.R2 are the two paired-end reads from one of the samples. For example, G2.R1.fq.gz, and G2.R2.fq.gz in the example data for the cv Grande Naine. Repeat this for all samples.

8. Prepare reference genome for mapping (*see* Note 20). The reference genome used for the banana example data can be found here https://www.ncbi.nlm.nih.gov/assembly/GCF_000313855.2. Click the download assembly button and select RefSeq and Genomic Fasta options.

9. Create a new directory for your analysis. For the test data, make a folder titled BananaGamma in the home directory. In this directory, create a new folder titled Genome. Place the .fna file downloaded in step 8 into this folder.

10. Index the genome for mapping. In the terminal window enter (cd) the Genome folder and execute the following command: bwa index *.fna. Five additional files will be produced.

11. Map the data using BWA-mem. Move (mv) the dedup.fastq.gz files created in step 7 into the project directory (*e.g.* BananaGamma). Execute the following command: bwa mem -M -t 4 ./Genome/*.fna Sample.R1.dedup.fastq.gz Sample.R2.dedup.fastq.gz > Sample.dedup.sam

This step will take many hours on a personal computer. The -t option sets the number of threads. If using Ubuntu, it may be helpful to launch the System Monitor software and select the Resources tab. This will graphically show the CPU usage and allow to monitor your computer to ensure it has not crashed.

12. Sort the sam file using samtools. In the terminal window, enter the following command:

Table 8.1 Partial output from bin-by-sam2.py using example data provided with this protocol

Chrom	Start	End	G2	N3	G2/G2	N3/G2
HE813975.1	1	100,000	107	122	3	3.446
HE813975.1	100,001	200,000	108	86	3	2.406
HE813975.1	200,001	300,000	181	114	3	1.903
HE813975.1	300,001	400,000	164	150	3	2.764
HE813975.1	400,001	500,000	150	135	3	2.72
HE813975.1	500,001	600,000	97	107	3	3.334
HE813975.1	600,001	700,000	107	93	3	2.627
HE813975.1	700,001	800,000	97	67	3	2.087
HE813975.1	800,001	900,000	97	88	3	2.742

samtools sort -O sam -T sample.sort -o G2_aln.sam G2.dedup.sam.
Where G2 is replaced with the sample name. Note that the output file name should end in _aln.sam for the bin-by-sam tool to work.

13. Convert SAM files to BAM format and index it for visual analysis in step 2 of Sect. 3.1.5. In the terminal window, enter the following command: samtools view -b G2_aln.sam > G2.bam

    Where G2 is replaced with the sample name. When complete, enter the following command:
    samtools index G2.bam. This will create an index file titled G2.bam.bai. Replace G2 with sample name.

14. Repeat steps 11–13 with all samples.
15. Install and run the bin-by-sam_2.0.py script. Download the Bin-by-Sam-tool into the sample processing folder (*e.g.* BananaGamma). Samples can either be compared to a wild-type reference, or to each other. Move the _aln.sam files created in step 12 to the Bin-by-Sam-tool folder. Open a terminal window, enter (cd) the directory and execute the following command: python bin-by-sam_2.0. py -o N3_100kbin.txt -s 100000 -b -p 3 -c G2_aln.sam. Where -o sets the output file, -s the bin size in base pairs (in the example data, a 100 kb bin size is used) -b inserts empty lines in the results table, -p sets the ploidy (3 for banana), and -c sets the cultivar control sample (sample G2 in the example data provided). When complete you should see the output file (in the example it is N3_100kbin.txt, for Novaria with 100 kb binning, change this name for different samples and binning), This folder can be opened with *e.g.* a spreadsheet to view and graph data (*see* Note 21 and Table 8.1).

3.1.5 Data Visualisation

1. Graph the data. The sample/control columns of the bin-by-sam output can be plotted as an Overlay Plot using a standard spreadsheet software such as

Microsoft Excel or LibreOffice Calc, or alternatives such as JMP or R. If using LibreOffice Calc (which comes preinstalled in Ubuntu), open the .txt file created in Sect. 3.1.4.15 step 15, select data from column G (the ratio of mutant to reference in the example) for one chromosome (chromosome 5 in the example data is labelled HE813979.1). Select Insert Chart from the drop-down menu. Select "Line Points Only" to produce a coverage graph (Fig. 8.3).

2. View data with IGV (optional). This tool provides a graphical view of mapped reads and can be a useful visual check of your mapping data. IGV can be used as a web app, which is preferred if the analysis computer has less than 16 Gb RAM. The genome file (.fna) from Sect. 3.1.4 step 8 needs to be renamed and indexed for IGV. Copy the .fna file to a new folder and change the extension from .fna to .fa. Next, open a terminal window, enter (cd) to the new folder and index by typing the following command: samtools faidx genome.fa. Where genome is the name of your genome file. Open a web browser and go to https://igv.org/app/. In the Genome pull down menu, go to the bottom (you may need to expand your browser to full screen in Ubuntu) and select Local File. Select both the .fa file and also the .fa.fai file that was created with samtools. Next, select Tracks, Local File to upload your bam files. This produces a graphical view of mapped reads (Fig. 8.4).

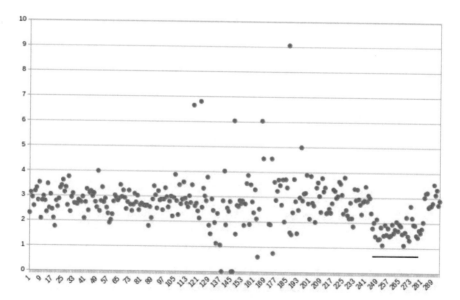

Fig. 8.3 Dosage plot analysis of chromosome 5 of mutant variety Novaria. Each dot represents a bin that is the mean coverage for 100 kb. Relative coverage values less than 3.0 indicate a putative deletion of one or more copies of a chromosome fragment while higher values (>3.0) indicate potential insertional events. The previously identified ~3.8 Mbp single copy deletion is underlined (Datta et al. 2018)

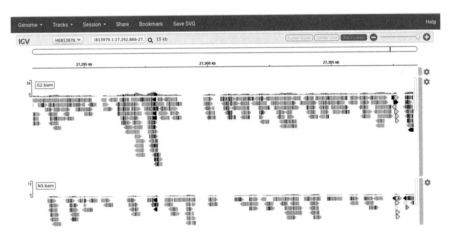

Fig. 8.4 Graphical view of mapped reads of cv Grande Naine (G2) and mutant variety Novaria (N3) example data using IGV

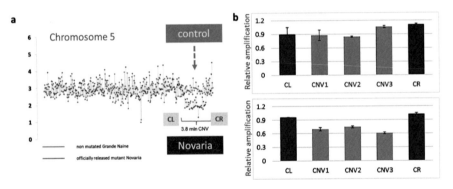

Fig. 8.5 (**a**) Stably inherited large deletion of 3.8 Mbp identified *via* LC-WGS in a 'Novaria' mutant. One hundred and eighty-nine genes are affected in the validated region by losing one copy. (**b**) qPCR verification of identified mutation (*CL* control left border, *CR* control right border, *CNV Region(s) showing deletion*). (Figure modified from Datta et al. 2018)

3.1.6 Validation of Predicted Variants

1. Select variants for validation. Review the bin-by-sam output table to select candidates and the graphed data as in Fig. 8.3.
2. Select regions falling within and outside the candidate regions from step 1.
3. Design primers for each of the regions, using *e.g.* Primer3, suitable for quantitative PCR (Rozen and Skaletsky 2000).
4. Perform quantitative PCR as outlined in Datta et al. (2018) for regions within and outside the candidate CNV. Variations in relative amplification abundance should correspond to the observed copy number change (Fig. 8.5).

4 Notes

1. While costly, tungsten carbide beads are durable and can be washed and re-used many times, making the actual cost per extraction extremely low. An alternative is to use 3-mm steel ball-bearings. Ball-bearings can be purchased in bulk at a low-cost allowing for disposal of bearings after a single use.

2. Ethidium bromide is mutagenic. Wear gloves, lab coat and goggles. Dispose of gloves in toxic trash when through. Avoid contaminating other lab items (equipment, phones, door handles, light switches) with ethidium bromide. Consult Material Safety Data Sheet (MSDS) for proper handling and disposal procedures. Alternative DNA stains, *e.g.* SYBR Safe DNA gel stain can be used instead of ethidium bromide but detection may vary with different dyes.

3. Lambda DNA is a high molecular weight DNA. Dilute this lambda stock in 1x TE to about 10 different concentrations in the range between 2.5 ng/μl and 150 ng/μl to serve as DNA concentration standards. The choice of the optimal DNA concentration standards depends on the concentrations of the sample DNAs. Calculate dilutions based on the information printed on the stock tube of the lambda DNA (note that the concentration of lambda DNA may vary from batch to batch). Store stocks at 4 °C.

4. Prepare gel loading dye containing 30% glycerol and color dye. Avoid loading dyes containing bromophenol blue or other dyes that migrate in a molecular weight range where you expect to observe DNA fragments. The presence of loading dyes can reduce the intensity of bands.

5. A variety of options exist for creating libraries compatible with Illumina sequencing-by-synthesis equipment. The per-sample library cost does not vary much at the time of writing this protocol, but cost savings can be achieved if library preparation will be routine and at a high scale as some suppliers sell library components individually. If using TruSeq, the kit comes with paramagnetic beads and all other components except for the adapter sequences (this may vary depending on the type of kit purchased). It is also common for sequencing service providers to provide library preparation services. Outsourcing the library preparation may be wise if you are new to the methods as the higher cost per library from a service provider will be balanced by the cost of new equipment and the time needed to gain expertise in your own laboratory.

6. All steps can be performed on a 64-bit computer with a minimum of 8Gb RAM. Mapping with BWA is computationally intense and can be very slow on computers with limited RAM. If necessary, mapping can be done on the cloud using a free option such as galaxy (https://usegalaxy.org/). If you are setting up a new computer for this work, we suggest a minimum of 16Gb RAM, 4 cores, and a free Linux operating system (for example, https://ubuntu.com/download/desktop). Guidelines for installing and running software in this protocol are written for Ubuntu 18.04 LTS.

7. Scripts have been developed to automate fastq processing and alignment (*see* http://comailab.genomecenter.ucdavis.edu/index.php/Bwa-doall) that are suitable for the bin-by-sam tool. Various tools have been updated since Bwa-doall was released and we have found that the python scripts require editing to properly work on Linux Ubuntu 18.04 LTS at the time of writing. Because of this, we provide alternative command line tools to remove duplicate reads and map data. Bash scripts can be written in Linux to automate the analysis steps if processing many samples. The bin-by-sam software version is no longer available at the original link. This link contains the version used in this protocol: (https://u.pcloud.link/publink/show?code=kZLR1xXZWNCbL3 m6HK7wRt50OvDfe8tGb9Mk).

8. The amount of tissue needed depends on the yield of DNA from the extraction procedure used and the method of sequencing library preparation. It is advised to test tissue collection and DNA extraction procedures in advance to ensure that the necessary quality and quantity parameters are met. It is advised to label collected samples with the line number, treatment number and generation. This is important in order to track the inheritance of induced mutations in tissue culture should material harbor chimeric sectors in the generation tested.

9. It is advised to store extracted DNA in a buffered solution (*e.g.* 10 mM Tris pH 8). Many DNA extraction kits come with an elution buffer containing Tris buffer. DNA can also be eluted/suspended in Tris-EDTA buffer as a buffer exchange is typically carried out in the first steps of library preparation.

10. Genomic DNA should be free of RNA and not degraded. In addition to fluorometry, it is recommended that DNA is evaluated by electrophoresis.

11. See Huynh et al. (2017) for a detailed description on the use of lambda DNA standards and image analysis for DNA quantification.

12. Use the Qubit concentration data to choose a volume and concentration of genomic DNA to produce a band with a pixel density within the range of the DNA standards used.

13. The DNA sample (genomic DNA band) should be completely out of the well and into the gel at least 2 cm. Do not run the gel too long as the genomic DNA band will become diffuse and hard to quantify. You may be able to accurately quantify degraded samples by altering electrophoresis conditions.

14. The required size range of DNA fragments and concentration of fragmented input DNA will vary depending on the library preparation method, sequencing chemistry and read length used. The original method to identify gamma-induced indels in banana was optimized for 2×300PE (paired-end) Illumina MiSeq sequencing with library preparation using a low-input Illumina "nano" kit. These kits enabled a low input of 200 ng by utilizing post-ligation PCR amplification to produce sufficient library for sequencing using Illumina sequencing-by-synthesis. A variety of kits available from different companies are suitable for discovery of gamma induced indels using this protocol. The 2×300PE sequencing chemistry supports 550 bp fragments. Note that newer sequencing platforms (*e.g.* NextSeq and NovaSeq) can produce higher throughputs at lower cost using shorter read lengths. If outsourcing sequencing, the

lowest cost option (in terms of raw Gb per dollar) will be suitable and fragment size and library preparation parameters can be adjusted for this. For the purposes of this protocol, parameters for low input libraries with 2×300PE sequencing are described. Different methods are available for DNA fragmentation. It is best to optimize fragment size utilizing a high sensitivity DNA system (*e.g.* Fragment Analyzer), however fragment sizes can be estimated using gel electrophoresis. Illumina 2×300PE sequencing-by-synthesis were used to prepare this protocol. However, higher throughput, shorter read sequencing provides a lower-cost alternative if sequencing is being outsourced.

15. Sequencing libraries are normalized using molarities so that the number of DNA molecules, independent of size or weight, can be adjusted for the sequencing run.

16. Libraries are typically pooled prior to sequencing because a sequencing run (*e.g.* flow cell lane) produces much more data (in base pairs) than is needed for a single sample. Pool samples at an equal concentration such that enough coverage is produced for each. For example, for a 660 Mb genome, if 10× coverage is desired, 6.6 Gb of sequence is needed per sample. If a sequencing run will produce 100 Gb of raw data, then 15 (100/6.6) samples can be pooled provided that quantification and pooling are accurate so that each of the 15 samples are equally represented in the sequencing reaction.

17. If using an Illumina MiSeq, follow manufacturer's protocols for on-board cluster generation and sequencing. If sequencing is being outsourced, the sequencing service will provide library QC, concentration determination, any necessary dilution, cluster generation and sequencing according to the equipment used.

18. Sequence reads are provided as compressed fastq format files that contain both reads that passed quality filters and those that did not.

19. Demultiplexing and trimming is often carried out by the sequencing provider. Check with your provider to determine what is included and how your data has been processed.

20. Data analysis involves (1) processing the raw data (fastq files), (2) mapping the raw (fastq) sequencing data to a reference genome, (3) calculating the average read depth over a set interval, or bin and (4) displaying the read depth bins to identify regions that deviate from expected values. Different tools have been described for these steps (*e.g.* BWA, Bowtie) and algorithms are under constant improvement. Users are encouraged to evaluate these scripts and modify algorithms as desired.

21. The output file contains one row per selected bin size with one column showing the number of reads in each bin and another column with the calculated dosage relative to the control sample. It is advised to try different bin sizes as results will vary depending on quality of sequence and depth of coverage. A bin size of 100 kb was previously used to detect gamma induced indels in banana. Ploidy (p) should be set to 3 if analysing triploid banana. When evaluating data of triploids, a relative coverage value less than 3 indicates a deletion of one or more copies of a chromosomal region while values greater than three indicate

potential insertional events. Different thresholds can be applied to filter potential false positive signals. In previous work, a threshold of three consecutive bins showing the same trend (below or above 3) was applied. This filter can be applied to the table of data independent of the visualization methods. Such variants were experimentally validated.

Acknowledgments Authors wish to thank Dr. Bernhard Hofinger and Dr. Prateek Gupta for support in setting up the bioinformatic pipelines. Funding for this work was provided by the Food and Agriculture Organization of the United Nations and the International Atomic Energy Agency through their Joint FAO/IAEA Centre of Nuclear Techniques in Food and Agriculture. This work is part of IAEA Coordinated Research Project D22005.

References

Datta S, Jankowicz-Cieslak J, Nielen S, Ingelbrecht I, Till BJ (2018) Induction and recovery of copy number variation in banana through gamma irradiation and low coverage whole genome sequencing. Plant Biotechnol J 16:1644–1653

Henry IM, Zinkgraf MS, Groover AT, Comai L (2015) A system for dosage-based functional genomics in poplar. The Plant Cell 27(9):2370–2383. Online, tpc.15.00349

Huynh OA, Jankowicz-Cieslak J, Saraye B, Hofinger B, Till BJ (2017) Low-cost methods for DNA extraction and quantification. In: Jankowicz-Cieslak J, Tai T, Kumlehn J, Till B (eds) Biotechnologies for plant mutation breeding. Springer, Cham

Jankowicz-Cieslak J, Till BJ (2015) Forward and reverse genetics in crop breeding. In: Al-Khayri JM, Jain SM, Johnson DV (eds) Advances in plant breeding strategies: breeding, biotechnology and molecular tools. Springer, Cham, pp 215–240

Jankowicz-Cieslak J, Huynh OA, Brozynska M, Nakitandwe J, Till BJ (2012) Induction, rapid fixation and retention of mutations in vegetatively propagated banana. Plant Biotechnol J 10: 1056–1066

Jankowicz-Cieslak J, Goessnitzer F, Datta S, Viljoen A, Ingelbrecht I, Till JB (2021) Induced mutations for generating bananas resistant to Fusarium wilt tropical race 4 – symposium proceedings. In: Sivasankar S, Noel Ellis TH, Jankuloski L, Ingelbrecht I (eds) Mutation breeding, genetic diversity and crop adaptation to climate change. CABI, Oxfordshire

Krasileva KV, Vasquez-Gross HA, Howell T, Bailey P, Paraiso F, Clissold L et al (2017) Uncovering hidden variation in polyploid wheat. Proc Natl Acad Sci 114:E913–E921

Rozen S, Skaletsky H (2000) Primer3 on the WWW for general users and for biologist programmers. In: Krawetz S, Misener S (eds) Methods in molecular biology bioinformatics methods and protocols. Humana Press, Totowa, pp 365–386

Siegel JJ, Amon A (2012) New insights into the troubles of aneuploidy. Annu Rev Cell Dev Biol 28:189–214

Tan EH, Henry IM, Ravi M, Bradnam KR, Mandakova T, Marimuthu MP et al (2015) Catastrophic chromosomal restructuring during genome elimination in plants. eLife Sci 4:e06516

Tan E, Comai L, Henry I (2016) Chromosome dosage analysis in plants using whole genome sequencing. Bio-Protocol 6. https://doi.org/10.21769/BioProtoc.1854

Till BJ, Datta S, Jankowicz-Cieslak J (2018) TILLING: the next generation. Adv Biochem Eng Biotechnol 164:139–160

Chapter 9
A Protocol for Detection of Large Chromosome Variations in Banana Using Next Generation Sequencing

Catherine Breton, Alberto Cenci, Julie Sardos, Rachel Chase, Max Ruas, Mathieu Rouard, and Nicolas Roux

Abstract Core activities of genebank operations include the preservation of germplasm identity and maintenance of genetic integrity. Some organisms such as banana are maintained by tissue culture that can foster accumulation of somatic mutations and loss of genetic integrity. Such changes can be reflected in their genome structure and thus be revealed by sequencing methods. Here, we propose a protocol for the detection of large chromosomal gains and/or losses that was applied to *in vitro* banana accessions with different levels of ploidy. Mixoploidy was detected in triploid (3x) accessions with chromosomal regions being diploid (2x) and tetraploid (4x) and in diploid accessions (2x) where large deletions resulted in partial haploidy (1x). Such abnormal molecular karyotypes can potentially explain phenotypic aberrations observed in off type material. With the affordable cost of Next Generation Sequencing (NGS) technologies and the release of the presented bioinformatic pipeline, we aim to promote the application of this methodology as a routine operation for genebank management as an important step to monitor the genetic integrity of distributed material. Moreover, genebank users can be also empowered to apply the methodology and check the molecular karyotype of the ordered material.

Keywords Aneuploidy · Banana · Chromosomal variation · Musa spp. · Somaclonal variants · Genebanks · NGS

1 Introduction

Somaclonal variation describes random cellular changes in plants regenerated through tissue culture. It occurs in certain crops that undergo micropropagation, and has been recorded in different explant sources, from leaves and shoots, to

C. Breton (✉) · A. Cenci · J. Sardos · R. Chase · M. Ruas · M. Rouard · N. Roux (✉)
Alliance Bioversity International-CIAT, Parc Scientifique Agropolis II, Montpellier, France
e-mail: c.breton@cgiar.org; n.roux@cgiar.org

© The Author(s) 2022
J. Jankowicz-Cieslak, I. L. Ingelbrecht (eds.), *Efficient Screening Techniques to Identify Mutants with TR4 Resistance in Banana*,
https://doi.org/10.1007/978-3-662-64915-2_9

meristems and embryos. Banana (*Musa* spp.) is a clonal crop that can be conserved and multiplied *in vitro*. Somaclonal variations have been observed in banana after prolonged periods of *in vitro* culture and after intensive multiplication phases, both resulting in increased rates of subculturing for a given clone. Although somaclonal variation can result in advantageous mutations that can be useful for the genetic improvement of banana, it is undesirable in the context of micropropagation and plant conservation. This type of variation indeed is a problem for genebank managers, whose objectives are to maintain the genetic integrity of their collections for subsequent research and breeding purposes, thus preserving genetic resources for future generations.

The International *Musa* Germplasm Transit Centre (ITC), managed by the Alliance of Bioversity-CIAT and hosted at the Katholieke Universiteit Leuven in Belgium, is the world's largest collection of banana germplasm with more than 1600 accessions of cultivated and wild species of banana (Ruas et al. 2017; de Langhe et al. 2018). It ensures the long-term conservation of a wide banana genepool and supports germplasm distribution all over the world. Due to the vegetative mode of propagation, banana accessions are kept *in vitro* under slow-growth conditions and regenerated through tissue culture. Although stress during the *in vitro* process is minimized by optimized multiplication and growing conditions, somaclonal variations have been observed. To avoid the conservation, and distribution, of material that holds such variations, and therefore ensure the maintenance of the genetic integrity of the germplasm, the ITC has developed the Field Verification exercise. This exercise aims at monitoring the genetic integrity of its banana accessions and combines evidence from morphological and molecular characterization to determine genetic integrity. To do so, plantlets that were maintained *in vitro* for more than 10 years are sent to the *Musa* Genotyping Centre (MGC) and to a partner field collection, USDA-ARS. In the lab, leaves are analyzed by flow cytometry and SSR markers, while in the field, plants are grown, characterized and photo documented. Once all results are obtained after a year or two, a panel of taxonomists check the morphological and molecular data and compare them to known reference information for each accession. Given that plants that have undergone somaclonal variations are expected to change in morphology, from changes in color to obvious growth issues, the panel of experts then assesses whether the accession is True-To-Type (TTT) or Off-Type (OT) (Chase et al. 2014). A major limitation of this process is the amount of time that is necessary to grow the plants and document them. It is also cumbersome to request the availability and expertise of key experts on a voluntary basis.

To fasten the process, the development of early screening methods is therefore of great interest for the community. To overcome the plant phenotyping bottleneck, investigation of modifications at the genome level has been targeted (Sahijram et al. 2003; Oh et al. 2007). The molecular basis of somaclonal variation is not precisely known, but both genetic and epigenetic mechanisms have been proposed (Kaeppler et al. 2000). Somaclonal variants in plants can be the result of various types of mutations such point mutations, gene duplication, transposable elements activation or large chromosomal rearrangements (in number and structure) (Bairu et al. 2011).

Current advances in Next Generation Sequencing (NGS) technologies and associated Genotyping-by-Sequencing methods allow the generation of high-throughput genetic markers for a large number of samples in a fast and cost-effective way. These datasets can therefore be used to study many events, including large chromosomal variations as recently reported in wheat and barley (Keilwagen et al. 2019). Such chromosome changes (e.g. aneuploidy) have also recently been reported in bananas (Cenci et al. 2019, 2021; Baurens et al. 2019, 2020). Although not all these variations may be the cause of off-type phenotypes, likely due to polyploidy that can mitigate them, it remains a change in DNA integrity of the plant that should be flagged, as well as a good reason for accession regeneration from backup or reintroduction with new material. These large chromosomal variations will be the focus of this chapter.

The protocol described here provides an early detection method for large chromosomal indels applied to material which can be obtained at ITC. This method could be used as a routine operation to check the genetic integrity of germplasm conserved *in vitro* in genebanks, and also in tissue culture laboratories. It can also be used with other methods for the quick testing of variations in mutants.

2 Materials

2.1 Plant Material

Passport data of accessions held at the ITC are published on the *Musa* Germplasm Information System (MGIS; www.crop-diversity.org). Available material from the ITC collection can be ordered at no cost for research, education and breeding purposes using a straightforward workflow (Fig. 9.1).

The MGIS website offers several options to search for germplasm that presents specific criteria such as taxonomy, country of origin, and it is also possible to identify accessions that have been evaluated for pest and diseases resistance, biotic or abiotic stresses and genome composition. To do so:

1. Go to https://www.crop-diversity.org/mgis/accession-search.
2. First refine your search by selecting "Yes" with the exchange availability filter.
3. Use the other filters to refine your search based on other criteria.
4. Click on "Add to my list icon".

MGIS permits users to create a list of accessions that, once registered and if the material is available for distribution, can be ordered online.

5. Go to "My list" on the top right menu and double check the list selected accession (s).
6. Click on the "order Germplasm material" button.

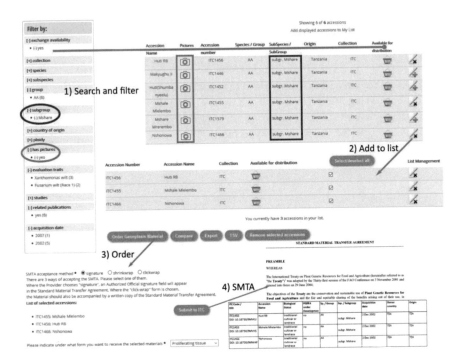

Fig. 9.1 Workflow to order banana samples from the International *Musa* Germplasm Transit Centre (ITC) *via* the *Musa* Germplasm Information System (MGIS)

7. If this is your first time on the website, create an account and indicate an accurate address to ensure the proper delivery of the plants.
8. Fill in the order form.

The *Musa* Online Ordering System (MOOS) is a three-step process which generates at the end of the process a Standard Material Transfer Agreement (SMTA) in PDF format that is automatically sent to the curator of the ITC collection for preparation of the material.

9. Select the appropriate SMTA acceptance method:

   (a) Signature: the document should be printed and signed by a person authorized to sign on behalf of the organization then sent to ITC for counter signature.
   (b) Shrink-wrap: by accepting the parcel, the Recipient is accepting the terms of the SMTA attached.
   (c) Clickwrap: you accept the SMTA online by clicking. It works the same way as when you order goods or services from other internet web sites.

10. Please indicate under what form you want to receive the materials.

Here we recommend selecting lyophilized leaf tissue which is the fastest way to receive it if your interest is DNA extraction for GBS or RADseq. Fresh leaves, that can be obtained by growing rooted plantlets, are recommended for whole genome sequencing.

11. Select the level of payment (article 6.7 or 6.11).

Although the material is distributed free of charge, several options exist to Enhance Benefit-Sharing if done for commercial purposes.

12. Indicate the purpose of the use.
13. Select your affiliation, professional and shipping address.
14. Indicate whether an import permit is needed.
15. Submit.

2.2 Data Generation

To detect changes in chromosome ploidy, SNP ratio changes need to be monitored along the chromosomes. It is therefore necessary to choose a technology that will generate high-throughput SNP-based markers for genome-wide marker discovery. Technologies based on restriction enzyme-mediated genome complexity reduction such as Genotyping by Sequencing (GBS) (Elshire et al. 2011), Restriction Site Associated DNA Sequencing (RADseq) (Davey et al. 2010), Diversity Arrays Technology Sequencing (DArTseq™) (Kilian et al. 2012) have advantages and disadvantages (Table 9.1) and are all appropriate to detect aneuploidy events. The choice of the technology can be influenced by existing datasets, in-house facilities or affordable solutions offered by service providers. Therefore, we don't provide a protocol for data generation but list the main points related to the main methods previously tested for such analyses.

3 Methods

The workflow described in Figs. 9.2 and 9.3 shows the different steps involved in the analysis. It is composed of two main processes. The aim of the first part is to map data DNA from DArTseq, GBS, RADseq and RNA from RNAseq onto a reference genome sequence (*Musa acuminata* genome (D'Hont et al. 2012; Martin et al. 2016), or *M. balbisiana* genome (Wang et al. 2019)), and perform a SNP calling. The second part of the pipeline uses the VCF file obtained from SNP calling to determine the genomic structure of the accessions and define their molecular karyotype in order to reveal possible ploidy change.

Table 9.1 NGS methods to generate high levels of polymorphism that can be used by the protocol

Molecular marker	Basis of polymorphism	Advantages	Disadvantages
GBS Genotyping by sequencing	Sequences of the ends of all resulting DNA restriction fragments produced by a frequent cutter enzyme; generates large no. of SNPs	Useful for high diversity & large genome species; cost effective for genomic-assisted breeding; high automation; technically easier to use and less demanding than RADseq	Proprietary technology
DArTseq Diversity arrays technology sequencing	Works on a genome complexity reduction concept – Selection of genome with predominantly active genes (target low copy sequences)	Reduction complexity methods are simple and cheaper than other GBS-based methods; high reproducibility; high heterozygotes representation	Single source for proprietary technology
RADseq Restriction site associated DNA sequencing	Sequences of short regions (50–150 bases) flanking each and all restrictions sites for a given endonuclease	Relatively low cost (greater no. of samples), and simple; greater coverage per locus; no prior genomic information required	Bias due to allele dropout, PCR duplicates and variance in depth of coverage among loci (all of the former vary according the RADseq method used)
RNAseq	Sequences of most of the genes in the genome, whole chromosome coverage except in centromeric regions	Studies may exist with such datasets	Expensive technique for looking only at genome structure. Transcription of alleles may vary across tissues and conditions

Step 1: Read cleaning, mapping, variant calling

- Read quality check step: Fastqc – Cutadapt
- Mapping reads on the reference: BWA – STAR
- Variant discovery: GATK

Step 2: Merge data, Use VcfHunter to establish the molecular karyotype

- Combine VCF sample
- Filtering
- Split VCF by chromosome
- Generate figure and assignation to each genome

The preprocessing pipeline presented in Fig. 9.2 shows the different steps that are compatible with four different sequencing technologies as listed in the previous section. The pipeline has to be done sample per sample according to the best practice of the software GATK (McKenna et al. 2010) used for the SNP calling part.

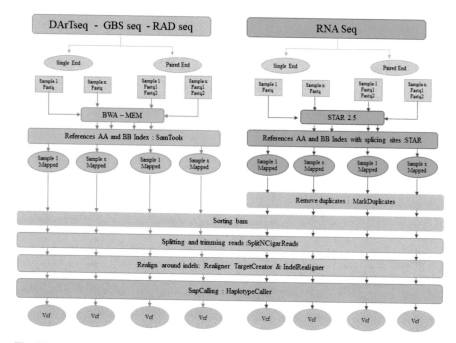

Fig. 9.2 Schematic overview of the bioinformatics workflow for SNP calling

Step 1: Read cleaning, mapping, variant calling

This pre-processing step performs the cleaning data to the SNP calling *via* the read mapping and should be performed on each individual in order to produce a single VCF by sample (to be combined in Step 2).

Read Quality Check

- Quality reads: Control the quality of the raw Fastq file with FASTQC

Description: In order to verify the quality of data reads, FastQC allows to check the quality score of each base and the presence or absence of the adaptor used to build the library. Adaptor depends on the sequencing technology used. https://www. bioinformatics.babraham.ac.uk/projects/fastqc/

- Trim low-quality base with Cutadapt

Description: According to the result of FastQC, Cutadapt (Martin 2011) trims low quality ends and removes adapters (Illumina).
Website: http://cutadapt.readthedocs.io/en/stable/guide.html

Fixed Parameters

-b AGATCGGAAGAGC (universal sequence for Illumina). Sequence of an adapter that was ligated to the 5′ or 3′ end. The adapter itself is trimmed and anything that follows too if located at 3′ end.

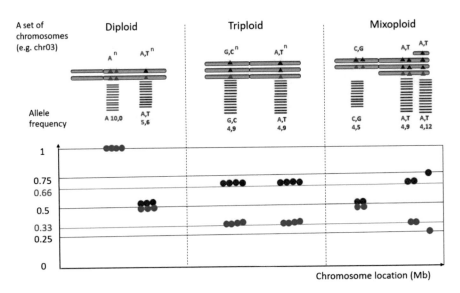

Fig. 9.3 SNP ratio calculation and visualization. Reads are mapped to the reference genome. For a given DNA base position, multiple reads will be aligned for a global coverage. SNP detected are assigned to a different color (corresponding to different genomes for hybrid species). SNP Frequency is calculated at each site (e.g. 0.5 = half of the reads display this allele) and then plotted on a graph according to their physical position along the chromosome. Variation of SNP frequency combined with SNP coverage along the chromosome indicate chromosome segment with ploidy change

-O 7: Minimum overlap length. If the overlap between the read and the adapter is shorter than LENGTH, the read is not modified. This reduces the no. of bases trimmed purely due to short random adapter matches.

-m 30: Discard trimmed reads that are shorter than 30.

--quality-base = 64: Assume that quality values are encoded as ascii (quality + QUALITY_BASE). The default (33) is usually correct, except for reads produced by some versions of the Illumina pipeline, where this should be set to 64. (Default: 33).

-q 20,20: Trim low-quality bases from 5′ and/or 3′ ends of reads before adapter removal. If one value is given, only the 3′ end is trimmed. If two comma-separated cutoffs are given, the 5′ end is trimmed with the first cutoff, the 3′ end with the second. The algorithm is the same as the one used by BWA (see documentation).

The tool generates a trimmed fastq file (*_cutadapt.fastq.gz) files for each accession.

Mapping Reads on the Reference

Description: Align reads on a reference genome (e.g. *Musa acuminata 'DH Pahang'*), with BWA (Li and Durbin 2010) for DNA, and STAR (Dobin et al. 2013) for RNA.

DNA Data: Map with BWA with default parameters with BWA-MEM.

Different types of genomic data such as DArTSeq, GBS and RADseq can be used (Fig. 9.2).

The tool generates a sam (*_.sam) files for each accession.

Website: http://bio-bwa.sourceforge.net/

RNA Data: Mapping with STAR in 2-pass mode.

Description: In the 2-pass mapping job, STAR will map the reads twice. In the first pass, the novel junctions will be detected and inserted into the genome indices. In the second pass, all reads will be re-mapped using annotated (from the GTF file given by the user) and novel (detected in the first pass) junctions. While this procedure doubles the run-time, it significantly increases sensitivity to novel splice junctions. In the absence of annotations, this option is strongly recommended.

The tool generates a folder for each accession, (names filled in column 3 "genome_name") filled in the configuration file, which contained the SAM files (converted in BAM file) of aligned reads and a .final.out file of mapping statistics for each library. In addition, a (--prefix) folder containing a mapping statistics file (--prefix + mapping.tab) for all accession is generated.

Website: https://github.com/alexdobin/STAR

Variant Discovery

- Add read group and (accession name from the fq.gz filename) sort BAM with Picard Tools

Description: This step replaces the reads Groups which describe the reads mapped on the reference, the sequencing technology, samples names, and library number are added.

ID = Read Group identifier (e.g. FLOWCELL1.LANE1).
PU = Platform Unit (e.g. FLOWCELL1.LANE1.UNIT1).
SM = Sample (e.g. DAD).
PL = Platform technology used to produce the read (e.g. ILLUMINA).
LB = DNA library identifier (e.g. LIB-DAD-1).

Website: https://broadinstitute.github.io/picard/

The tool generates a bam (*_rmdup.bam) file for each accession with the RG (Read Group) modified.

- Mark duplicate reads and index BAM with MarkDuplicates from PicardTools

Description: PCR duplicate removal, where PCR duplicates arise from multiple PCR products from the same template molecule binding on the flow cell. These are removed because they can lead to false positive variant calls. Sort the BAM file and mark the duplicated reads.

Website: https://broadinstitute.github.io/picard/

The tool generate a bam (*_rmdup.bam) files for each accession with duplicated reads removed. In addition, a file named (--prefix + rmdup*stat.tab) file containing duplicate statistics for each accession was generated in the (--prefix) folder.

- Index BAM with Samtools

Description: This step reorders the bam file according to the genome index position.

Website: http://samtools.sourceforge.net/

The tool generates a reordered (*_reorder.bam) and bai (*_reorder.bai) files for each accession.

- Split 'N CIGAR' reads with SplitNCigarReads from GATK

Description: Splits reads that contain Ns in their cigar string (e.g. spanning splicing events in RNAseq data). Identifies all N cigar elements and creates k+1 new reads (where k is the number of N cigar elements). The first read includes the bases that are to the left of the first N element, while the part of the read that is to the right of the N (including the Ns) is hard clipped and so on for the rest of the new reads. Used for post-processing RNA reads aligned against the full reference.
Website: https://gatk.broadinstitute.org/hc/en-us/articles/360036727811-SplitNCigarReads

The tool generates a split and trimmed (on splicing sites) bam (*_trim.bam) and bai (*_trim.bai) files for each accession.

- Realign indels with IndelRealigners from GATK (2 steps)

Description: The mapper BWA has some difficulties to manage the alignment close to the Indel. The step is not necessary with HaplotypeCaller but is necessary with UnifiedGenotyper. The tool generates a bam (*_realigned.bam) and bai (*_realigned.bai) files realigned around indel for each accession. This step is done with the GATK version 3.8.
Website: https://github.com/broadinstitute/gatk-docs/blob/master/gatk3-tutorials/(howto)_Perform_local_realignment_around_indels.md

The tool generates a bam (*_realigned.bam) and bai (*_realigned.bai) files realigned around indel for each accession.

- Create a VCF file with HaplotypeCaller from GATK

Description: The HaplotypeCaller is able to call SNPs and indels simultaneously *via* local de-novo assembly of haplotypes in an active region. Whenever the program encounters a region showing variation, it discards the existing mapping information and completely reassembles the reads in that region. This allows the HaplotypeCaller to be more accurate when calling regions that are traditionally difficult to call.

Parameters:

--genotyping_mode DISCOVERY.
--variant_index_type LINEAR.
--variant_index_parameter 128,000.

https://gatk.broadinstitute.org/hc/en-us/articles/360036712151-HaplotypeCaller

Note: All these steps can be performed separately or with workflows reported in the literature (e.g. Toggle(Monat et al. 2015; Tranchant-Dubreuil et al. 2018)) or using the scripts we made available on GitHub at https://github.com/CathyBreton/Genomic_Evolution, which follow all the steps.

The tools are developed in Perl, bash, Python3, Java and work on the Linux system and require:

- Bamtools v2.4, https://github.com/pezmaster31/bamtools
- BWA v0.7.12, http://bio-bwa.sourceforge.net
- STAR v2.5, https://github.com/alexdobin/STAR
- GATK v4, https://software.broadinstitute.org/gatk/
- Picard Tools v2.7, https://broadinstitute.github.io/picard/
- Samtools v1.2, https://github.com/samtools/samtools
- VCFHunter v1, https://github.com/SouthGreenPlatform/VcfHunter

Description: Get one FASTQ file ready for SNP calling per accession from raw sequence data (fastq.gz files).

> **USAGE:**
> <Technique>_<Type_of_data>_fastq_to_vcf_job_array_Total_GATK4.pl *-r ref. fasta -x fq.gz -cu accession*
> Parameters:

-r (string): reference FASTA filename.
-x (string): file extension (fq.gz).
-cu (string): cultivar.

Step 2: Molecular Karyotyping and Ploidy Change Detection

Based on known sequence variability, SNP variants can be assigned to the ancestral genomes in order to plot the genome allele coverage ratio and to calculate the normalized site coverage along chromosomes as described in (Baurens et al. 2019). This method can detect chromosome changes such as homoeologous exchanges (Cenci et al. 2021) but also is powerful enough to detect ploidy variation along the chromosomes as illustrated in Fig. 9.3.

The method has been developed within the VCFHunter software (Baurens et al. 2019) and can be used with the following procedure (Fig. 9.4).

- Merge datasets

Script name: CombineVcf.pl.

Description: The script merges and prepares the final VCF file, this step combines multiple VCF and performs pre-filtering using GATK. The samples to analyze are combined to the reference samples to allow allele assignation. Reference samples are representative genotypes that are relevant to the identification of your samples (i.e. *acuminata* or *balbisiana* without admixture). Whenever necessary, such SNP datasets can be downloaded on MGIS (https://www.crop-diversity.org/mgis/gigwa with RADseq_ABB_AB datasets) *via* GIGWA (Cenci et al. 2021; Sempéré et al. 2019).

> **USAGE:** perl CombineVcf.pl -r reference_fasta -p output_prefix -x extension_file_to_treat.

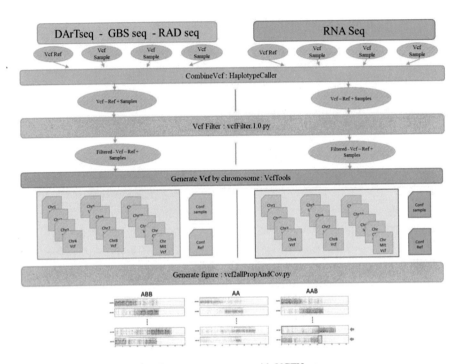

Fig. 9.4 Pipeline to determine the genome structure with VCFHunter

Parameters:

-r (string): reference FASTA filename.
-p (string): VCF output prefix.
-x (string): file extension (VCF).
Website: https://github.com/CathyBreton/Genomic_Evolution

- Filter SNP dataset

Script name: vcfFilter.1.0.py

Description: Filter VCF file based on most common parameters such as the coverage, missing data, MAF (minor allele frequency). The tool keeps bi-allelic sites and removes mono-allelic, tri-allelic, tetra-allelic sites.

USAGE: python3 vcfFilter.1.0.py --vcf file.prefiltered.vcf --prefix file.filtered --MinCov 8 --MaxCov 200 --MinAl 3 --MinFreq 0.05 --nMiss 50 --names All_names. tab --RmAlAlt 1:3:4:5:6:7:8:9:10 --RmType SnpCluster.
Website: https://github.com/SouthGreenPlatform/VcfHunter/blob/master/tutorial_ DnaSeqVariantCalling.md

Parameters:

--MinCov, --MaxCov (int): min and max coverage for each genotype. If not, converted into missing data.
--MinAl (int): min coverage for each allele. If not, converted into missing data.
--MinFreq (float): minor allele frequency.

--nMiss (int): max number of accession with missing data--RmAlAlt : keeping diallelic sites.

--RmType SnpCluster (string): remove SNP clusters (define as 2 adjacent SNPs).

--names (string): A file containing accession names (one per line) in the output file.

• Split VCF by chromosome

Description: Generate a VCF file for each chromosome with VcfTools (Danecek et al. 2011), in order to obtain a representation (chromosome Painting) of the SNP position along each chromosome.

    USAGE: vcftools --vcf file_filtered.vcf --chr chr01 --recode --out batchall_filt_chr01.

    Parameters:

--chr (string): generate a VCF from a given chromosome.

--recode: generate a new VCF file.

--out (string): output file name.

• Generate molecular karyotype

Description: The program allows to perform a chromosome painting for all chromosomes of a given accession.

    Script name: vcf2allPropAndCov.py.

    USAGE: python3 vcf2allPropAndCov.py --conf <chromosomes.conf> --origin <origin.conf> --acc <sample_name> --ploidy 2

    Parameters:

--acc: name of the sample to be analyzed (as in the VCF file).

--conf chromosomes.conf: list of VCF files for the chromosomes to be processed.

--origin origin.conf: name of the sample to be used for allele attribution.

--ploidy: expected ploidy of the sample (e.g. 2, 3 or 4).

The tool generates the 4 following files:

_AlleleOriginAndRatio.tab is a file describing each grouped allele, its origin and the proportion of reads having this allele at the studied position in the accession.

_stats.tab is a file reporting statistics on SNP sites used, sites where an allele is attributed to each group and alleles number attributed to each group in the accession.

_Cov.png is a figure showing read coverage along chromosomes (see figure below for interpretation).

_Ratio.png: is a figure showing grouped allele read ratio along chromosomes (see figure below for interpretation).

Website: https://github.com/SouthGreenPlatform/VcfHunter/blob/master/tutorial_ChromosomePainting.md

Use Cases: Application to Accessions at ITC

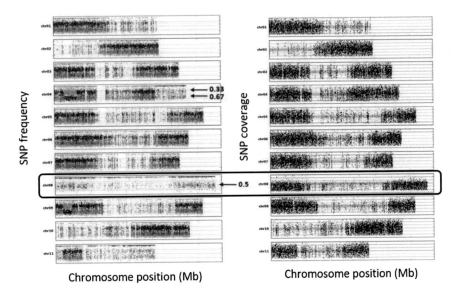

Fig. 9.5 Molecular karyotype of an ABB banana cultivar. On the left, SNP ratio distribution along the chromosomes. Each SNP is illustrated by a dot and assigned to a genome by a color (A = green, B = red). The red arrows with values (0.33/0.5/0.67) refer to the SNP frequency ratio. On the right, read coverage at SNP positions along the chromosomes. Heterozygous SNP frequency distribution around 0.5 and lower SNP coverage along the whole chromosome 8 indicate a BB pattern (loss of chromosome 8A) in 'Dole' (ITC0767)

This section describes several examples of samples across the banana taxonomy that were processed by our method and allowed the detection of large chromosome variations.

Chromosomal changes in allotriploids (AAB, ABB).

Comprehensive characterization of the ABB samples has been conducted on ITC materials, revealing the genome structure or molecular karyotypes of most of the existing taxonomic subgroups (Cenci et al. 2021). Using VCFHunter on RADseq data, we were able to uncover patterns of chromosome segment recombinations between A and B genomes for most of the accessions. Among them, one displayed a clear case of chromosome number change as illustrated on Fig. 9.5. For this genotype belonging to the Bluggoe subgroup, no SNP variant was assigned to the A subgenome (in green) along the whole chromosome 8. Most of the B variants (in red) were located at the top of the diagram (value = 1) with an unimodal distribution for residual B genome heterozygosity around 0.5 (instead of 0.33/0.67 expected in the presence of three B chromosome 8). Moreover, the diagram on the right shows a lower SNP density coverage in comparison to all the other chromosomes, showing that chromosome 8 was diploid (2x) for this accession. This pattern observed here is due to the loss of the A genome version of chromosome 8. For comparison, chromosome 4 exhibits two regions with irregular patterns compared to

a

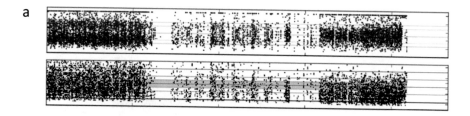

b

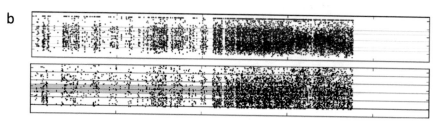

Fig. 9.6 Patterns of chromosome loss and gain in *Musa* ABB (A = blue, B = red, read coverage = black). (**a**) Pattern of 'Simili Radjah' (ITC0123) chromosome 5, showing frequency of A and B variants (y axes) around 0.5 and lower SNP coverage on the second arm of one B chromosome, indicates AB pattern (loss of one B chromosome 5 second arm). (**b**) Pattern of 'INIVIT PB-2003' (ITC1600) chromosome 10, containing an interstitial region showing frequency of A and B variants (y axes) around 0.5 and higher read coverage indicates AABB pattern (duplication of A chromosome interstitial region)

expectation for an ABB. One is located on the first arm of the chromosome and the second one is placed on the distal part of the second arm. However, in both regions, allelic frequencies for heterozygous sites are consistent with triploidy (0.33/0.67) and no distortion of the coverage density is observed.

Two additional examples of aneuploidy detection in ABB are illustrated in Fig. 9.6. In the accession 'Simili Radjah' (ITC0123, ABB), the loss of the second arm in one of the B chromosomes of chromosome 5 can be inferred by the SNP frequency at 0.5 for both A and B assigned SNP and by the lower SNP density coverage compared to the first arm (Fig. 9.6a). In the accession 'INIVIT PF-2003' (ITC1600, ABB), in the second arm of chromosome 10, an interstitial region appears to have SNP frequency at 0.5. Since the SNP density coverage in this region is higher than in the remaining chromosome having 1A and 2B chromosomes, the duplication of the interstitial region in chromosome A was deducted, being the SNP ratio in this region 2A:2B.

Finally, other examples were also detected in AAB Plantain cultivars (Fig. 9.7). We observed in 'Nzumoigne' (ITC0718), a SNP frequency of A and B alleles at 0.5 on chromosome 2 that was combined to lower SNP density coverage (compared to other chromosomes as exemplified with chromosome 3) (Fig. 9.7a). In 'Ihitisim' (ITC0121), multiple events were detected including a chromosome gain on chromosome 3 with SNP ratio of ~0.25 and ~ 0.75, supported by a higher SNP coverage and a partial loss at the beginning of the chromosome 4 (Fig. 9.7b).

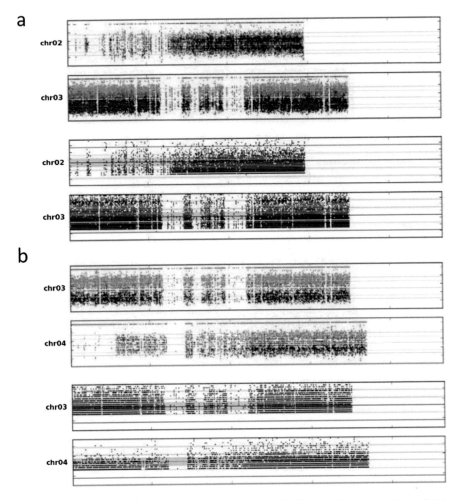

Fig. 9.7 Patterns of chromosome loss and gain in *Musa* AAB Plantains (A = green, B = red). (**a**) Pattern of 'Nzumoigne' (ITC0718) chromosome 3 with regular pattern (variants A and B with frequency at 0.67 and 0.33 (y axis), respectively) compared to chromosome 2 having both ends with A and B variants at 0.5 frequency. The pattern indicates loss of both ends of one A chromosome 2. (**b**) A and B variant frequencies (y axes) in chromosome 3 (0.75 and 0.25, respectively) and read coverage higher in chromosome 3 than in chromosome 4 indicate the presence of an additional chromosome 3A in 'Ihitisim' (ITC0121)

Chromosomal changes in diploid banana accessions (AA).

The survey of more than 200 AA accessions with the same procedure revealed molecular karyotypes corresponding to chromosome arms with large interstitial or terminal deletions for a few individuals (Fig. 9.8).

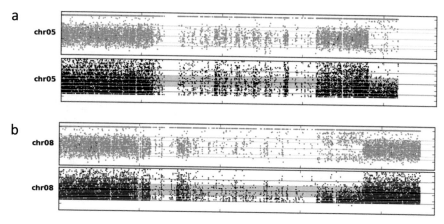

Fig. 9.8 Chromosome pattern of mutated chromosomes in cultivated AA diploid accessions. (**a**) Patterns of 'No.110' (AA, ITC0413) second arm terminal region of chromosome 5. Absence of heterozygous variants and lower read coverage indicates terminal deletion in one copy of second arm. (**b**) Patterns of 'Pahang' (AA, ITC0727) second arm interstitial region of chromosome 8. Absence of heterozygous variants and lower read coverage indicates interstitial deletion of a copy on second arm

4 Notes

1. Lyophilized leaf tissue allows sufficient DNA quality extraction for GBS, RADseq or DArtSeq (Doyle et al. 1990).
2. We recommend using the PSTI restriction enzyme for GBS, RADseq or DArTseq (Chan et al. 2014).
3. From our experience, GBS will generate fewer markers than RADseq but higher read coverage (Hueber et al. 2015).
4. A minimum read coverage of 10x by haplotype would be required to support SNP detection statistically (e.g. 30x for triploids).
5. SNP calling by individual and then merge is more efficient for SNP calling. However, SNP calling can be performed in different way and used directly with Step 2 on molecular karyotyping (https://github.com/CathyBreton/Genomic_Evolution).
6. The pipeline provides graphical output for interpretation of insertion deletions by chromosome. A workflow is developed in github (https://github.com/CathyBreton/Genomic_Evolution).
7. The method can detect aneuploidy. In case of partial gain on a chromosome, as this is based on allelic ratio mapped of the reference genome, the type of event such as insertion or duplication cannot be distinguished. Inversion and translocation are not identified. (Cenci et al. 2019)
8. Large chromosome variations provide information on the loss of genetic integrity. However, those events in banana seems to not be systematically synonyms of somaclonal variations leading to change in observed phenotype. Further research is needed to clarify the importance of the events linked to the loss of True-To-Typeness status.

Acknowledgements This work was supported by the CGIAR Fund, and in particular by the CGIAR Research Program Roots, Tubers and Bananas.

References

Bairu MW, Aremu AO, Van Staden J (2011) Somaclonal variation in plants: causes and detection methods. Plant Growth Regul 63:147–173. https://doi.org/10.1007/s10725-010-9554-x

Baurens F-C, Martin G, Hervouet C, Salmon F, Yohomé D, Ricci S et al (2019) Recombination and large structural variations shape interspecific edible bananas genomes. Mol Biol Evol 36:97–111. https://doi.org/10.1093/molbev/msy199

Busche M, Pucker B, Viehöver P, Weisshaar B, Stracke R (2020) Genome sequencing of Musa acuminata dwarf Cavendish reveals a duplication of a large segment of chromosome 2. G3: genes, genomes. Genetics 10:37–42. https://doi.org/10.1534/g3.119.400847

Cenci A, Hueber Y, Zorrilla-Fontanesi Y, van Wesemael J, Kissel E, Gislard M et al (2019) Effect of paleopolyploidy and allopolyploidy on gene expression in banana. BMC Genomics 20:244. https://doi.org/10.1186/s12864-019-5618-0

Cenci A, Sardos J, Hueber Y, Martin G, Breton C, Roux N, Swennen R, Carpentier SC, Rouard M (2021) Unravelling the complex story of intergenomic recombination in ABB allotriploid bananas, Ann Bot 127(1):7–20, https://doi.org/10.1093/aob/mcaa032

Chan A, Xavier Perrier, Christophe Jenny, Jean-Pierre Jacquemoud-Collet, Mathieu Rouard, Julie Sardos, Nicolas Roux, Christopher D Town (2014) Using genotyping-by-sequencing to understand Musa diversity. Poster 449 PAG XXIII Conference. San Diego (USA)

Chase R, Sardos J, Ruas M, Van den Houwe I, Roux N, Hribova E, et al (2014) The field verification activity: a cooperative approach to the management of the global Musa in vitro collection at the international transit Centre. In: XXIX international horticultural congress on horticulture: sustaining lives, livelihoods and landscapes (IHC2014): IX 1114, pp 61–66

D'Hont A, Denoeud F, Aury J-M, Baurens F-C, Carreel F, Garsmeur O et al (2012) The banana (Musa acuminata) genome and the evolution of monocotyledonous plants. Nature 488:213

Danecek P, Auton A, Abecasis G, Albers CA, Banks E, DePristo MA et al (2011) The variant call format and VCFtools. Bioinformatics 27:2156–2158. https://doi.org/10.1093/bioinformatics/btr330

Davey JW, Davey JL, Blaxter ML, Blaxter MW (2010) RADSeq: next-generation population genetics. Brief Funct Genomics 9:416–423. https://doi.org/10.1093/bfgp/elq031

de Langhe E, Laliberte B, Chase R, Domaingue R, Horry JP, Karamura D, et al. The 2018 The 2016 Global strategy for the conservation and use of *Musa* genetic resources – key strategic elements. Acta Hortic 71–78. doi:https://doi.org/10.17660/ActaHortic.2018.1196.8

Dobin A, Davis CA, Schlesinger F, Drenkow J, Zaleski C, Jha S et al (2013) STAR: ultrafast universal RNA-seq aligner. Bioinformatics 29:15–21. https://doi.org/10.1093/bioinformatics/bts635

Doyle JJ, Doyle JL (1990) A rapid total DNA preparation procedure for fresh plant tissue. Focus: 13–15

Elshire RJ, Glaubitz JC, Sun Q, Poland JA, Kawamoto K, Buckler ES et al (2011) A robust, simple genotyping-by-sequencing (GBS) approach for high diversity species. PLoS One 6:e19379. https://doi.org/10.1371/journal.pone.0019379

Hueber Y, Sardos J, Hribova E, Van den Houwe I, Roux N, Rouard, M (2015) Application of NGS-generated SNP data to complex crops studies: the example of Musa spp. (banana). Poster presented at Plant and Animal Genome - PAG XXIII Conference. San Diego (USA) 10–14

Kaeppler SM, Kaeppler HF, Rhee Y (2000) Epigenetic aspects of somaclonal variation in plants. Plant Mol Biol 43:179–188. https://doi.org/10.1023/A:1006423110134

Keilwagen J, Lehnert H, Berner T, Beier S, Scholz U, Himmelbach A et al (2019) Detecting large chromosomal modifications using short read data from genotyping-by-sequencing. Front Plant Sci 10. https://doi.org/10.3389/fpls.2019.01133

Kilian A, Wenzl P, Huttner E, Carling J, Xia L, Blois H et al (2012) Diversity arrays technology: a generic genome profiling technology on open platforms. Methods Mol Biol 888:67–89. https://doi.org/10.1007/978-1-61779-870-2_5

Li H, Durbin R (2010) Fast and accurate long-read alignment with burrows-wheeler transform. Bioinformatics 26:589–595. https://doi.org/10.1093/bioinformatics/btp698

Martin M (2011) Cutadapt removes adapter sequences from high-throughput sequencing reads. EMBnetjournal 17:10. https://doi.org/10.14806/ej.17.1.200

Martin G, Baurens F-C, Droc G, Rouard M, Cenci A, Kilian A et al (2016) Improvement of the banana "Musa acuminata" reference sequence using NGS data and semi-automated bioinformatics methods. BMC Genomics 17:243. https://doi.org/10.1186/s12864-016-2579-4

McKenna A, Hanna M, Banks E, Sivachenko A, Cibulskis K, Kernytsky A et al (2010) The genome analysis toolkit: a MapReduce framework for analyzing next-generation DNA sequencing data. Genome Res 20:1297–1303. https://doi.org/10.1101/gr.107524.110

Monat C, Tranchant-Dubreuil C, Kougbeadjo A, Farcy C, Ortega-Abboud E, Amanzougarene S et al (2015) TOGGLE: toolbox for generic NGS analyses. BMC Bioinf 16:374. https://doi.org/10.1186/s12859-015-0795-6

Oh TJ, Cullis MA, Kunert K, Engelborghs I, Swennen R, Cullis CA (2007) Genomic changes associated with somaclonal variation in banana (Musa spp.). Physiol Plant 129:766–774. https://doi.org/10.1111/j.1399-3054.2007.00858.x

Ruas M, Guignon V, Sempere G, Sardos J, Hueber Y, Duvergey H et al (2017) MGIS: managing banana (Musa spp.) genetic resources information and high-throughput genotyping data. Database (Oxford) 2017. https://doi.org/10.1093/database/bax046

Sahijram L, Soneji JR, Bollamma K (2003) Analyzing somaclonal variation in micropropagated bananas (Musa spp.). In Vitro Cell Dev Biol Plant 39:551–556

Sempéré G, Pétel A, Rouard M, Frouin J, Hueber Y, De Bellis F et al (2019) Gigwa v2—extended and improved genotype investigator. Gigascience 8. https://doi.org/10.1093/gigascience/giz051

Tranchant-Dubreuil C, Ravel S, Monat C, Sarah G, Diallo A, Helou L et al (2018) TOGGLe, a flexible framework for easily building complex workflows and performing robust large-scale NGS analyses. bioRxiv:245480. https://doi.org/10.1101/245480

Wang Z, Miao H, Liu J, Xu B, Yao X, Xu C et al (2019) Musa balbisiana genome reveals subgenome evolution and functional divergence. Nature Plants 1. https://doi.org/10.1038/s41477-019-0452-6

The opinions expressed in this chapter are those of the author(s) and do not necessarily reflect the views of the IAEA: International Atomic Energy Agency, its Board of Directors, or the countries they represent

Part IV
Low-cost *In Vitro* Methods for Banana Micropropagation

Chapter 10
Low-Cost *In Vitro* Options for Banana Mutation Breeding

Babita Jhurree-Dussoruth

Abstract Mutation fixation of irradiated banana cultures is achieved through at least three generation advancements by *in vitro* subculturing. The *in vitro* culture is a technique which allows rapid multiplication of plantlets within a short time and which often relies highly on expensive inputs that are almost unaffordable in many developing countries. This chapter highlights some easily affordable options that can be adopted for *in vitro* propagation and weaning of tissue-cultured banana plantlets and other horticultural crops. The presented options provide resource restricted laboratories opportunities to coopperate with irradiation facilities for mutation induction. Thus, when applied to any locally selected banana variety, the low-cost *in vitro* methods allow an efficient mutagenesis process to improve local accessions. Low-cost alternatives adopted to carry out *in vitro* mutagenesis activities in the current FAO/IAEA project are presented, by using as baseline other cheaper options developed and adapted through a locally funded project (supported by the Mauritius Research and Innovation Council).

Keywords Tissue-culture · Low-cost · *In vitro* mutagenesis · Banana · Mutation breeding

1 Introduction

Banana breeding of popular varieties through conventional methods is a long and tedious process, challenged by low female fertility, polyploidy and heterozygosity (Jenny et al. 2002; Ortiz 2013). Unlike seeded species, the classical cross-breeding for genetic improvement of the mostly sterile and seedless banana can last over several years. This has prompted many countries to have recourse to mutation breeding as an alternative method for the improvement of banana.

B. Jhurree-Dussoruth (✉)
Food and Agricultural Research and Extension Institute, Reduit, Mauritius
e-mail: bdussoruth@farei.mu

© The Author(s) 2022
J. Jankowicz-Cieslak, I. L. Ingelbrecht (eds.), *Efficient Screening Techniques to Identify Mutants with TR4 Resistance in Banana*,
https://doi.org/10.1007/978-3-662-64915-2_10

151

Earliest studies on the effect of radiation on plant development dates back to 1928 by Lewis Stadler (Stadler 1928; FAO/IAEA 2011) with groundwork on induced mutation breeding of banana starting in the early 1970's by De Guzman, Ubalde and Del Rosario (De Guzman et al. 1976). Since then, significant advancement has been achieved in mutation breeding of banana through use of physical mutagens (X- or gamma rays) or chemical mutagens (ethyl methanesulphonate, EMS) (FAO/IAEA 1995). Successful applications of mutation in genetic improvement of banana include the development of an early flowering mutant of banana cv. Grande Naine, named 'Novaria' by F. J. Novak (Mak et al. 1996) and an improved local variety named "Al Beely" banana (FAO/IAEA 2011). Furthermore, Jain et al. (2011) reported on ten banana mutants/variants that were identified as promising with improved traits such as earliness, improved bunch and fruit size, reduced plant height or tolerance to Fusarium wilt.

Countries facing important banana pest and disease pressures can have access to improved genotypes such as FHIA hybrids developed by the Honduran Foundation for Agricultural Research, however, many countries still opt to improve their popular locally adapted variety. Mauritius is one such country which embarked on mutation breeding of its popular banana varieties through the joint support of the FAO and IAEA. A first project aimed at inducing tolerance/resistance to Panama disease in a highly prized local dessert type banana, namely 'Gingeli' banana (AAB genome) from year 2004 to 2015 using gamma rays on *in vitro* cultures. In the absence of hotspots, the generated mutant lines were screened at greenhouse level and the surviving lines were multiplied for field testing. Till date, no promising line has been identified. In another FAO/IAEA funded project, mutagenic treatments were performed on three other popular varieties, the 'Mamoul' (ABB), 'Ollier' (AAA) and D. Cavendish banana (AAA) with aim to induce tolerance to Fusarium wilt race 1 (in 'Mamoul') and race 4 in the others.

In vitro mutation breeding is a powerful tool for induction of desired traits. However, the restricted access to irradiation and *in vitro* facilities remains among the main withdrawal factors for these countries to embark on banana breeding using plant biotechnology techniques. The Plant and Breeding Genetics laboratory of the FAO/IAEA in Seibersdorf, Austria is one such laboratory that supports developing countries by irradiating plant materials, free-of-charge. Similarly, Mauritius can also extend physical mutagenic treatments to neighboring countries.

In vitro culture methods are used in banana mutation breeding whereby emerging buds from the mutagenized cultures are isolated and brought back into culture, through a series of up to four subcultures (generations) (M_1V_0 to M_1V_4) which then allows the mutation to be fixed in the genome (*in vitro* mutagenesis) (Roux 2004). The *in vitro* tissue culture (TC) is often associated with high production costs as it relies on aseptic conditions, high quality chemicals, high-energy and expensive glassware which restrict its broad application in plant breeding. To make TC affordable without compromising the quality of the plants, several low-cost alterna-tive components of tissue culture have been considered and adopted (Kodym and Zapata-Arias 2001; FAO/IAEA 2004; Ogero et al. 2012; Datta et al. 2017).

A minimum, recommended number of shoot tips (500–1000) can be irradiated and if managed in batches, the number of plants generated by the M_1V_4 population can be easily handled using the low-cost approach. The sections below highlight the low-budget approach of *in vitro* banana mutation breeding.

1.1 Low-Budget In Vitro Options

1.1.1 Infrastructure

A typical TC laboratory consists of rooms dedicated to each aspect of *in vitro* culture: A first room (usually the first entry point) is for the reception and handling of mother plants received from the field. The other rooms include the media preparation room, the transfer room with laminar flow cabinets which serve as an aseptic area for handling of *in vitro* cultures and a growth room with controlled temperature and light for growth of cultures. In a low-cost situation, a simple large room or two-rooms can be partitioned to accommodate the above areas while respecting the flow of each section and ensuring that the aseptic area is against the flow of air.

1.1.2 Consumables and Equipment

Banana cultures are mainly established in Murashige and Skoog (MS media) or modified MS media supplemented with auxin and cytokinin. The media can be prepared by using powdered form of ready-made MS media or by mixing aliquots of four stock solutions of macronutrients, micronutrients, iron and vitamins (*see* Annex 10.1). The latter is more accessible and relatively cheaper for developing countries. High quality analytical grade chemicals are expensive and not a requisite for routine TC. On the other hand, commercial grade chemicals of lower purity can be easily procured from school suppliers that provide such chemicals as teaching materials and their use can lead to up to 70% savings (Dussoruth et al. unpublished). In a study, Gitonga et al. (2010) reduced the cost of producing tissue culture banana seedlings by 93.9% by using alternative nutrient sources. Similarly, Ogero et al. (2012) successfully realized cost reduction of over 80–95% by substituting conventional MS compounds by locally available fertilizers (for macro/micronutrients), over-the-counter vitamins (for vitamins) and seaweeds or agricultural plant hormones (for plant growth regulators, PGR's). However, Dussoruth et al. (unpublished) found that the quality of agricultural fertilizers had significant effect on *in vitro* plants. Aliquots of field fertilizers used to make the MS media did not dissolve well, leaving behind deposits of impurities that reduced the effective percentage of each element in the media. Moreover, the explants inoculated in these media died due to phytotoxicity. On the other hand, MS media made up with hydroponic salts did not cause phytotoxicity but significantly reduced the multiplication rate of banana plants. Thus, all substitutions that aim in cost reduction need to be fine-tuned as these can affect the quantity and quality of the plants.

Gelling agents and high-grade sucrose constitute the most expensive components of the culture media (Sahu and Sahu 2013). The gelling agents account for over 70% of media costs while the media chemicals contribute to 5–15% of production costs (Prakash 1993). Dussoruth et al. (unpublished) also noted that when both high-quality chemicals and analytical grade sucrose are used for MS media, sucrose becomes the largest components of MS media (63–78%) and this is reduced to about 3–8% when replaced by household sugarcane-based sugar (see Table 10.1). On the other hand, when commercial grade chemicals and household sugar are used, the gelling agents account to approximately 70–80% of media cost (Dussoruth et al. unpublished). As the media chemicals (nutrients and PGR's) contribute to only about 10–20% of overall cost of 1 L media (*see* Table 10.1), a significant cost reduction in media is possible mainly by substituting the expensive high-quality gelling agents and analytical sucrose by cheaper commercial substitutes. It is now standard practice by many laboratories to use commercial sugar in media preparation.

Banana cultures can be generated in solid, semi-solid or liquid media. The liquid media allows a more rapid proliferation (Ahloowalia and Prakash 2004) but relies on expensive orbital shaker at around 60–80 rpm for media aeration to prevent asphyxia of the explants. The orbital shaker can, however, be replaced by a locally mounted shaker. In the absence of a shaker, the cultures can be multiplied in stationary liquid media by using supports as interface (matrix), which serve as an intermediate layer between the liquid and explants. Cheaper matrices such as marbles (beads), paper shreds, glass wool, filter paper bridges, cotton fibers among others were used by several researchers as reported by Prakash et al. (2004) and Datta et al. (2017). Dussoruth et al. (unpublished) assessed the effectiveness of sugarcane based bagasse, cotton wool, paper shreds, tissue-paper, marbles (*see* Fig. 10.1) for micropropagation of banana and African violet (an ornamental, *Saintpaulia ionantha*). Tissue-paper and cotton wool were as effective as phytagel or agar and agitated liquid, but the disadvantage was that explants, including the roots, were strongly attached to the matrix. The size of the marbles are important because large spaces among the beads can lead to drowning of explants.

As shakers/agitators add to production costs, many laboratories opt for semi-solid or solid media by using gelling agents which contribute to the viscosity of the

Table 10.1 Percentage cost contribution of each component in MS media depending on quality of chemicals and quality of gelling agents used

Media component	Commercial grade chemicals (%)	Analytical grade chemicals (%)
Chemicals and PGR	8–21	9–12
Sucrose source	3–8	63–78
Gelling agent-Agar or Phytagel	71–89	10–28

Fig. 10.1 Use of low-cost physical support matrices for growth of cultures. The matrices (from left: cotton, cotton, tea paper, marble, liquid-agitated, bagasse) and the banana cultures, 4 weeks after subculture

medium. The solidified media acts as an interface allowing the explants to be seated in the media while uptaking necessary nutrients for growth and development. Conventional gelling agents are agar, agarose, and gellan gum (marketed under trade names such as phytagel or gelrite) (Prakash et al. 2004). The gellan gums produce transparent media, allow easy detection of contamination and are of higher quality than agar, however they are often expensive. In Mauritius, the gelrite and phytagel cost around 1200–1600 €/kg and 500–700 €/kg respectively while agar costs 90–400 €/kg, depending on the purity. To reduce the cost of media, cheaper sources of phytagel can be mixed with agar. Agar is the most widely used gelling agent, since it is usually unnecessary for high purity agar in large-scale micropropagation (Prakash et al. 2004). Extensive research has been done into cheaper alternatives to replace agar and gellan gums or in combination with other gelling agents (Puchooa et al. 1999; Wilson and Tenkouano 2020 and as compiled by Prakash et al. 2004; Datta et al. 2017).

Dussoruth et al. (unpublished) tested household tapioca pearls (locally called as 'sagoo'), rice flour, cornstarch and arrowroot as alternative gelling agents, of which rice and tapioca were least suitable. Arrowroot gel led to improved plant development compared to agar. However, the 100% cornstarch-based or arrowroot-based media is firm and the opaque white to grayish-white color make difficult to detect any contamination (*see* Fig. 10.2). Addition of agar (1:1 ratio) softened the media while allowing explant growth and development. To allow detection of contamination during the initial stages, it is advisable to avoid the opaque media and use mainly agar as gelling agent. An agar or phytagel-based medium can be cooked in a microwave to produce a homogeneous medium prior to dispensing in culture jars, while the household gelling agents need to be carefully cooked in a pan on a stove with continuous stirring, to avoid lumps.

Fig. 10.2 Effects of different low-cost gelling agents and physical matrices. Effect of (from left to the right) agar, arrow root, cornstarch, arrowroot plus agar and cornstarch plus agar on banana shoot proliferation. Addition of agar to the low-cost gelling agents improved the structure of the media

The protocol in 'Methodology' section describes the banana micropropagation on modified MS media solidified with agar-cornflour (1:1) and agar-arrowroot (1:1). Growth of cultures was comparable to those grown in agar-based media (*see* Fig. 10.2). Cost of cornstarch and arrowroot was around 4 €/kg which is only 0.1% of the cost of agar. As 80–100 g of the cornstarch and arrowroot is used per litre of MS media compared to 2.5 g/l phytagel and 8.0 g/l agar, this represented a respective 60% and 80% savings over agar and phytagel.

Water is another main component which is used to make stock solutions and the culture media. Conventionally, distilled, doubled distilled or de-ionized water is used (Ahloowalia and Prakash 2004). These are expensive and can further significantly increase the cost of production if they are operated using electricity. As reported by Ahloowalia and Prakash (2004), several researchers used cheaper options such as operating distillatory with gas or altogether replacing distilled/deionized water with cheaper alternatives like tap, rain or bottled water. Experiments using distilled/deionized, tap, bottled and household-filtered water showed that cultures were successfully produced using filtered water (*see* Fig. 10.3). Filtered water was then adopted as the most economic source for preparation of both media and stock solutions used in this project. The unit cost of a home-scale filtration unit ranged from € 50 to € 125, depending on brand and quality compared to about € 5000 for a deionizer/distillatory unit leading to a cost reduction of over 95%.

Big commercial laboratories use high quality glassware that tend to be costly and fragile. Similarly, chemicals or stock solutions which are accurately measured during media preparation using expensive equipment can be approximate to the

Fig. 10.3 Effect of five water sources (from left to the right tap, bottled, household-scale filtered, deionized and double-distilled) used in media preparation. Performance of banana cultures in media made using filtered water was comparable to those from deionized water

nearest amount. Over 90% cost savings can be made by using easily available, hardy and cheap wares such as syringes (for dispensing on aliquots of 1 ml to 50 ml) or plastic cylinders for larger volumes. These can easily be procured over the counter from school supply shops or medical stores. Access to culture vessels, namely 'Magenta' commonly used for TC may not be easily found in developing countries, alternatively recycled jam jars can be used. Cheap polypropylene (PP) lids can be used to replace the metal caps with added advantage that they do not rust and allow light to reach the cultures. Additionally, cling film or sterile PP plastic films can also be placed around the mouth of the jar and held tight using rubber bands (*see* Fig. 10.4). This method is not recommended, as wrapping the cover/film around the mouth of the jar takes more time than closing with a lid. It however, remains an option for those not having access to required lids. The Guangdong Academy of Agricultural Sciences (GAAS) uses a special PP pouch (*see* Fig. 10.4) for growing cultures both under natural and artificial LED lights.

While most activities can be easily handled using cheaper alternatives, some steps such as pH testing, require a closer monitoring. Some laboratories used pH indicator paper as low-cost option, however this method is not very accurate (Ahloowalia and Prakash 2004) because slight changes in pH are not easily detected through the indicator paper. Alternatively, portable hand-held cheaper pH meters can be used and they cost about 10% of analytical high accuracy pH meters. The expensive magnetic stirrers (€ 300 to € 900) can equally be substituted by simple ones at less than € 50.

An aseptic condition is a major requisite in any TC activity. Wares and media are best sterilized using autoclave, which can be costly. Commercial laboratories working with large volumes can afford autoclaves but this can be an expensive option for laboratories with small turn-over rates. Stericlave (portable medical pressure steam

Fig. 10.4 Options for
culture vessels and covers.
From left: jar covered with
polypropylene (PP) cap,
conical flask covered with
an aluminium and jam jar
with metal cap

Fig. 10.5 (**a**) Stericlave. (**b**) Pressure cooker. (**c**) Low-cost, glass hood

sterilizer) of above 40 l, which cost about 30% of a conventional autoclave is an affordable option (*see* Fig. 10.5a). Similarly a pressure cooker (20–25 l volume) which cost about 6–10% of an average autoclave can also be used (*see* Fig. 10.5b). Effective steam sterilization can be achieved through automatic setting of autoclaves at a pressure of 15 psi (1.05 kg/sq.cm) along with a temperature of 121 °C for at least 30 min. On the other hand, the pressure cooker needs closer monitoring as overcooking can lead to caramelization of the medium (see 'Methods' section). Another point to consider is that a filled pressure cooker can be very heavy to lift and thus the height of the table, where the stove will be placed to heat the cooker, may need to be adjusted for easy handling. A pressure cooker is an easily available option for small laboratories where for example in a 20 l capacity pressure cooker, only about thirty 200–250 ml capacity culture (jam) jars can be stacked. As each jam jar is filled with about 30 ml medium, a total of only 1 l media can be sterilized at a time in a 20 l capacity cooker.

In vitro culture manipulation is done inside a laminar hood in a transfer room under aseptic conditions. Instead of using high purity ethanol, commercial ethanol can be equally used to clean working surface and a flame to sterilize the scalpels and forceps. In the absence of glass beads sterilizer, the sterilization can easily be done by dipping the working ends of the tools in alcohol and then immediately flaming them over an alcohol lamp or gas burner. A laminar flow cabinet allows handling of cultures in a sterile environment and the cheapest ones varies around 5000–6000 €. If the laboratory intends to undertake TC works for many years, then it is advisable to invest in a laminar hood. Otherwise, if the hood is needed only to allow generation advancement of the mutated population, a glass hood can be designed such that the opening is wide enough only to allow easy manual handling of the cultures. The glass hood can be cuboidal about 1 m wide, 40 cm deep and 60 cm high with an opening on one side (of about 30–40 cm from base) (*see* Fig. 10.5c). The hood should be placed against flow of air and surface sterilized regularly.

After placing the explant in the media, the jar is immediately closed with the cover to prevent microbial contamination. A layer of parafilm is often additionally wrapped around the base of the cover to ensure that the entry point for any microbes is blocked. The parafilm, however, is relatively expensive and can be easily substituted by equally effective food wrap (cling films), which thus allows a 90% savings over parafilm.

1.1.3 Light and Energy

The cultures are conventionally placed on shelves illuminated with cool light fluorescent tubes (conventionally at 1200–2000 lux) (Ahloowalia and Prakash 2004) under a 16 h:8 h light/dark cycle at 25 ± 2 °C. Artificial lighting of cultures accounts to nearly 60% production cost (Ahloowalia and Savangikar 2004). However, many *in vitro* growing plants can tolerate wide fluctuations with temperatures (Ahloowalia and Savangikar 2004) as high as 30 °C and as low as 10 °C with improved plant growth (Kodym et al. 2001). Research works have also been reported on the performance of *in vitro* cultures under natural light using solatube (Kodym et al. 2001), nethouse, light-emitting diodes (LEDs) (Datta et al. 2017), domelight or diffused light (Ahloowalia and Savangikar 2004) and sidewise lighting system rather than downward illumination of culture racks (Datta et al. 2017).

In this project, all cultures were grown under natural light from either a solatube (purchased at around 1200 €) (*see* Fig. 10.6) or from a 1 m^2 domelight made of superior quality acrylic (purchased at around 120 €), that were fixed on the roof of a 9 m^2 room. In a separate study, cultures were grown in a netcloth shed that was lined with plastic running from the roof to the sides (*see* Fig. 10.6), and also in a room, receiving diffused light through the windows.

Alternative culture growing conditions are summarized in Table 10.2. The solatube redirects the daylight through reflective tubing without heating the room and can illuminate an area of 3–5 m^2 (Kodym et al. 2001) whereas the domelight can heat up the room, requiring the window to be opened for air flow.

Depending on the season, the number of buds produced per explant in diffused light was reduced or comparable to conventional controlled conditions. The growth

Fig. 10.6 Alternative sources of natural light that can be potentially used for micropropagation. Left image: Shed, lined up with plastic and netcloth, which can be rolled up depending on heat accumulation. Right image: Solatube directs diffused light into the room, without heating the room

Table 10.2 Comparison of radiation and temperature of the different environment in Mauritus

	Illumination on sunny day Lux[a]	Minimum Temperature, °C	Maximum Temperature, °C	Prevalent average temperature, °C
Outside	53,000–70,000	Not measured		
Room with solatube	3000–4000	22 – summer	32 – summer	20–30
		18 – winter	27 – winter	
Room with domelight	3000–4500	24 – summer	44 – summer	20–40
		23 – winter	40 – winter	
Shed (plastic on top and side, side also with white net)	4000–6000	22 – summer	51 – summer	Variable
		15 – winter	45 – winter	
Room with diffused light	400–900 (depending on time of day)	21	28	25

[a]Measured using LIG1050 lux meter

rate was also slower under diffused light. Conditions with the solatube were more stable, while the domelight's conditions were directly dependent on the weather. Proliferation rates were better in rooms lit by the solatube than by the domelight. While no electricity is involved in use of natural light, each system has its own advantages and constraints and need to be optimized prior to application (*see* Fig. 10.7).

In the shed, the rate of development was not uniform. Mortality was relatively high (50–80%) depending on the outside conditions. The afternoon sun, especially in summer, led to intense scorching due to hot microclimate within the jars. This can be reduced by covering the west side of shed with shade netting. The other problem

Fig. 10.7 Response of banana cultures under different growing conditions, from left to the right: control, diffused light (2 plants), shed. Those grown under diffused light were comparable to control, but with slower rate. In shed, mortality and scorching were high. Remaining explants developed but with rate of development depending on the season. Development of rooted plantlets was also possible under diffused light

associated with growing cultures in a shed was the accumulation of dusts which intensified risks of contamination when brought back in the transfer room. However, cultures at rooting stage can be grown in such conditions as the plants can be sent to nursery for hardening.

In a commercial micropropagation laboratory, where the volume and quality of planting materials to be produced are crucial, there is often little concern about using cheaper options. The above low-cost substitutes have been proposed for countries with restricted or no TC facilities so that they can proceed with *in vitro* mutagenesis to improve any local banana cultivar for which conventional breeding can be too lengthy or complex. The low-cost alternatives can as well be included in a routine TC where up to 80% cost reduction can be achieved (*see* Table 10.3).

1.2   In Vitro Mutagenesis

In vitro methods as described by Vuylsteke (1998) and Lopez et al. (2017), can be applied to mass multiply the desired local accessions to generate enough explants that can then be sent to another laboratory for irradiation. This section describes only the physical mutagen (gamma rays) which was applied on excised banana shoot tips.

Table 10.3 Cost savings through adoption of low-cost substitutes of the TC components

Component	Conventional input	Low-cost substitute	Percentage (%) cost savings
Chemical and PGR		Use of commercial grade	70%
Gelling agents	Agar or Phytagel	Cornstarch or arrow root	60% (when used in combination with agar, 1:1) 85–90% when used alone. Reduction can be higher, in case higher purity agar or gelrite are used.
Sucrose	Analytical, high purity grade	Tea/household sugar	95–99%
Culture vessels, Glasswares	- Magenta vessels - Pyrex grade volumetric flasks - Measuring cylinders	- Recycled glass jars - Plastic graduated measuring cylinders, syringes	80–98%
Water source	Electrical double distillator	Household/semi-commercial water filtration system	90–95%
Sterilization	Autoclave	Stericlave	60–70%
		Pressure cooker (20 to 25 l)	90%
Sterile transfer room	Laminar hood	Hood made of glass (non-electric)	98%
Growth rooms	Fluorescent tubes/ air-conditioner (for a 9 m^2 room electricity can cost about € 1200–1400/year)	- Solatube (for a 9–16 m^2 room) - Domelight (for a 9–16 m^2 room) - Diffused light (from windows)	- With solatube investment cost covered in 1 year - Almost no electricity needed with other options
Weaning	Automated misting, wet wall for first stage hardening to maintain temperature and humidity	Simple mounted structures	Over 80–90%

In order to reduce risks of chimerism, the explants are trimmed to produce shoot-tips of about 2–3 mm length.

After irradiation, the mutation is fixed in the genome by carrying out three subcultures (the whole process is referred to as generation advancement). In order to dissolve potential chimeric sectors during each subculture, longitudinal sections are made through the explant to select buds or propagules (*see* Fig. 10.8). The mother explant can be further subcultured to recover the main meristem mutation after homohiston formation.

Fig. 10.8 Explant with buds after incubation ready for generation advancement. The arrows indicate buds/propagules that will be carried over in the next generation after subculture. Longitudinal sections are made such that the propagules with part of mother tissue is transferred to fresh media

In practice, however, it is often difficult to dissociate the buds from the mother explant. In such cases, routine subculturing/decapitating the main explant is practiced to promote shoot proliferation. After irradiation, the cultures are referred to as the M_1V_0 generation and the buds thus produced (after 4–6 weeks of incubation) are subcultured and make the population of M_1V_1 individuals and this process is repeated after each incubation period until the population of M_1V_3 is reached. The resultant M_1V_3 population is placed in regeneration medium (usually of half strength macronutrients and without any PGR) for the buds/shoot to regenerate into plantlets (or referred to as putative mutant line). These individual mutant lines might be clonal propagated for multiple traits screening, particularly abiotic or biotic stress which usually uses destructive methods. Moreover, the development of multiple clones per individual is also important as avoid loss of the mutant lines during the weaning of *ex vitro* plantlets.

1.3 Weaning of Mutant Lines

Prior to screening the mutant lines for any traits of interest, such as to Fusarium wilt using the double-tray system or field tests, the rooted plantlets are hardened so that they adapt to the outside environment where the light intensity (4000–12,000 lux) is higher compared to growth rooms that are artificially illuminated (1200–2000 lux) and the relative humidity lower (40–80% v/s 98–100%) (Ahloowalia and Savangikar 2004). Weaning is preferably done gradually, starting with high humidity and partial shade which is then gradually decreased. Commercial laboratories rely on greenhouses with sections of (i) partial shading (70–80%) and intermittent misting, (ii) 40–50% shade with regular irrigation/overhead fine irrigation and (iii) 20% shade to full sunlight with irrigation. In order to support developing countries, an easy method of weaning is proposed to handle the mutant lines in batches.

The high humidity in the weaning stage is created by placing the plantlets on a table and covering them with a frame, lined with clear polyethylene sheeting, and regularly watering the plants (or mist using a spray bottle) (*see* Fig. 10.9). The partial shade is provided by covering the structure with shadecloth (70–80% shade). The plastic sheet is removed after about 2–3 weeks to allow the plants to adjust to ambient relative humidity. The hardening media should be freely draining and relatively rich. Jhurree-Dussoruth and Kallydin (2011) successfully hardened banana plantlets in media consisting of sterile soil and manure (*see* Fig. 10.9). The addition of perlite improves the porosity of the medium and is important mostly in the first stage hardening.

In this chapter, a low-budget TC and weaning procedure are presented to allow laboratories with limited resources to proceed with banana mutation breeding after an outsourced irradiation. Banana cultures in this project were irradiated using local facilities, but an additional section has been included to guide those who need to seek an international support for irradiation.

Fig. 10.9 Weaning and hardening stages. (**a**) & (**b**) Simple structures constructed out of metal framework, lined with polyethylene sheeting. During the hardening, the sheet covers the whole structure and is gradually lifted after 2–3 weeks, to allow plantlets to adapt to ambient humidity. (**c**) Testing of most appropriate potting media from locally available materials. Sterile mix of manure and soil (1:1) allowed rapid development of plants 4 weeks after weaning. (**d**) Fully acclimatized plants ready for transfer to the experimental plot

2 Materials

2.1 Preparation of Media

1. MS stock solutions I to IV made as per Annex 10.1 (stored in refrigerator).
2. Gelling agents (technical grade agar, commercial grade cornstarch and arrowroot starch).
3. Granulated table sugar.
4. Jam jars (250 ml) with metal caps.
5. Stove, pan and ladle.
6. Pressure cooker/stericlave.
7. Top loading weighing balance (1- 200 g, (2 d.p)).
8. Portable pH meter, calibrated with dilute (0.1 N NaOH and 0.1 N HCl for pH adjustment).
9. Magnetic stirrer with magnets.
10. Plastic, graduated measuring cylinder (100 ml).
11. Graduated 2 l plastic beaker.
12. Syringes (2 ml, 5 ml, 25 ml or 50 ml).
13. Rod.
14. Kitchen jug for dispensing of cooked media.
15. Refrigerator.

2.2 Shoot-Tip Establishment

1. Suckers of desired variety.
2. Knife.
3. Detergent dish soap in water.
4. Wetting agent (Tween or Teepol).
5. Broad spectrum fungicide.
6. Commercial bleach (at 2%).
7. Sterile water.
8. Magnetic stirrer/magnets.
9. Plastic container (1 l).
10. Glass jar (250 ml) with caps.
11. Laminar air-flow cabinet or glass hood.
12. Household alcohol.
13. Alcohol lamp.
14. Cotton roll.
15. Scalpel and sterile blades.
16. Forceps, medium (20 cm) and long (30 cm).
17. Culture jar with low-cost gelling agent-based MS medium (supplemented with BAP at 5 mg/l or 2 mg/l).

18. Smooth, white/off-white ceramic tile (15 × 15 cm).
19. Growth Chamber.

2.3   Shoot Tip Multiplication/Generation Advancement

1. Scalpel and removable sterile blades.
2. Forceps: medium (about 20 cm) and long (about 30 cm).
3. Smooth, white/off-white/creamy ceramic tile (15 × 15 cm).
4. Recycled newspaper.
5. Aluminium foil.
6. Household alcohol.
7. Alcohol lamp.
8. Cotton roll.
9. Laminar air-flow cabinet or glass hood.
10. Growth chamber.
11. Culture jar with low-cost gelling agent based MS medium (supplemented with BAP at 5 mg/l).

2.4   Growing of Cultures

1. Shelves (meshed to allow natural light to reach lower shelves).
2. Naturally lit room (either 'solatube' or 'domelight').
3. Minimum/maximum thermometer.

2.5   Mutation Breeding/Generation Advancement

1. Banana proliferating tissues (raised from shoot-tip, after fourth subculture onwards).
2. Irradiated banana cultures (stage M_1V_0).

2.6   Rooting

1. As in above Sect. 2.3, except item no. 11.
2. Culture jar with low-cost gelling agent-based MS medium (no PGR).

2.7 Weaning

1. Seedling trays.
2. Beaker or bucket for washing off of gels.
3. Clean pair of scissors.
4. Perlite, peat, sterile commercial soil, sterile farmyard manure.
5. Fertiliser (slow release or complex fertiliser such as N/P/K, 13:13:20:2).
6. Potting bags (2 l).
7. Water for spraying.
8. Spray bottle.
9. Wooden or metal-frame structure (5 m × 2 m × 1 m high) lined with clear polythene sheet.
10. Shade cloth (70–80% and 40–50% shade) to be placed on roof of shed.
11. Bench to place trays or pots.

3 Methods

3.1 Preparation of Stock Solutions

1. Prepare stock solution of plant growth regulator (PGR) at the rate of 0.1 g/l water for each PGR: 6-Benzylaminopurine (BAP) and Indole-3-acetic acid (IAA) (*see* Note 1).
2. Refrigerate until use.

3.2 Preparation of Multiplication Medium

1. In a beaker, add 100 ml of stock I, 1 ml of stock II, 5 ml of stock III and IV.
2. Add 50 ml of BAP, 2 ml of IAA.
3. Add 500 ml filtered water, add 30 g sucrose, stir until sugar is dissolved.
4. Pour in 1 l measuring cylinder or leave graduated beaker. Make up to almost 1 l.
5. Adjust pH to 5.8.
6. Pour in saucepan, select any of the two gelling agents

(a) Agar/cornflour gelling agent: Add 40 g cornflour and 4 g agar (media are labelled as 'AC5' representing the 5 mg/l BAP)
(b) Agar/arrow root gelling agent: Add 40 g arrow root and 4 g agar (media are labelled as 'AA5' representing the 5 mg/l BAP)

7. Stir well until there are no clumps.
8. Cook on slow flame until a semi liquid consistency is obtained.

9. Dispense 30 ml of media in culture jars and close jars immediately.
10. Sterilise in:

   (a) Stericlave set at 121 °C at a pressure of 15 psi for 20–25 min
   (b) Pressure cooker as follows:

      – Place a perforated plate (on a stand, with gap of about 4–6 cm) inside cooker, add tap water just below the rim
      – Place jars on plate, cover and heat cooker on medium flame until first whistle
      – Leave on medium flame for 25–30 min after first whistle

11. Allow to cool down, remove from cooker, close caps tightly and store at room temperature until use.

3.3   Preparation of Rooting Medium

1. Proceed as above (Sect. 3.2), except in step 1 use only 50 ml aliquot of Stock I, and in step 2 no plant growth regulator is added. The media are labelled as AC0 or AA0, respectively for the agar/cornflour and agar/arrow root gelling agents (*see* Note 2).

3.4   Sterilization of Tools

1. Wrap one long forceps, one medium forceps and scalpel in aluminium foil.
2. Wrap working tile in newspaper and then with aluminium foil (to save on foil, newspaper is used).
3. Sterilize as above (Sect. 3.2, point 10).

3.5   Sterilization of Working Surface

Prior to all works under laminar or hood, spray household alcohol to whole working table and side of hood with cotton.

3.6   Establishing Banana Cultures

This section highlights steps from culture initiation until cultures reach stage for mutagenesis. Details can be obtained in Vuylsteke (1998).

3.6.1 Culture Initiation

1. Select and uproot suckers from elite mother plants (take care to avoid damaging base which contains the meristem).
2. Wash thoroughly outside the lab and using knife, remove outer leaves to produce 5–7 cm tall and 3 cm wide block of tissue resting on 3 cm tall corm (base) (now referred to as explants). Bring inside the laboratory.
3. Wash explant in running water with washing detergent for 10 min.
4. Using sharp knife, trim explant to smaller size (3 cm tall and 0.7 cm wide block) of tissue on 1 cm tall corm.
5. Place explants in a beaker of water containing a broad-spectrum fungicide (0.5%) and a few drops of wetting agent and allow stirring for 5 min (*see* Note 3).
6. Bring the beaker under sterile hood. Ensure dissecting tools, sterile working tile, dipping alcohol and flame are ready for regular sterilization of the dissecting instruments.
7. Surface disinfect the explants in diluted household bleach (2%) for 15 min with a few drops of wetting agent and swirl frequently.
8. Decant bleach and reduce explant to size of about 1.5 cm tall using scalpel and sterile medium forceps. The aim is to remove tissues damaged by bleach and excise inner tissues containing the meristem.
9. Carry out three rinses in sterile filtered water for 5, 10, 15 min respectively (swirl frequently inside hood).
10. Further trim explants to about 0.5–1 cm tall that contain the meristem with corm. Trimming is done on the sterile tile using sterile dissecting tools (*see* Note 4).
11. Transfer (using long forceps) to initiation MS-based medium supplemented with 5 mg/l BAP+ 0.175 mg/l IAA. Media are labelled as MS5 (*see* Note 5).
12. Transfer jars to growth room. Cover jars with black cover for 3–4 days until explant turn green.

3.6.2 Multiplication of Shoot-Tip Cultures

1. After 2–3 weeks, bring jar to Transfer Room. Trim the brown/black/oxidised parts and base and transfer the explant into fresh multiplication medium. Bigger explants can also be cut longitudinally through the apex. This is referred to as fist subculture (s1).
2. In first subcultures, no new buds will be seen. Buds often start to be produced after the second subculture. During subculture, the original explants can be decapitated if they are big. They can also be split longitudinally. At that stage it is not necessary to cut off each single bud/sprout and inoculate separately, they can be transferred as small clumps.
3. During repeated subculture, split the bigger explants longitudinally and subculture each part. Depending on multiplication rate, repeat subculturing every 5–7 weeks, to promote rapid multiplication (*see* Note 6).

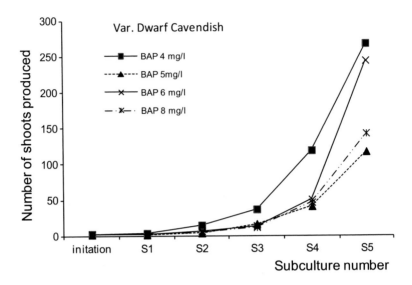

Fig. 10.10 Number of buds produced from explant after each subculture (4–6 weeks) in banana (var. Dwarf Cavendish, AAA) at different BAP levels. The explants entered the rapid phase of multiplication after the 4th subculture. Explants for irradiation were selected among the population of explants from the 4th or 5th subculture onwards

4. After repeated subculture of explants (often by 4th to 5th, depending on variety), the multiplication rate will increase rapidly, and explants will enter the log phase of multiplication (*see* Fig. 10.10). Select explants from these stages for irradiation.

3.6.3 Shoot-Tip for Mutagenesis – Dispatch and Reception

1. In order to reduce effects of chimerism in shoot-tip mutation breeding, the explants are trimmed to 4–5 mm length leaving as little vegetative tissue around the meristem as possible.
2. When preparing the bulk for irradiation, inoculate in small batches, as the trimmed explants can easily dehydrate if left on working surface during this process.
3. The explants are preferably inoculated on multiplication (solid) media (*see* Note 7).
4. It is advisable to carry out a radiosensitivity test to determine the optimum dose for bulk irradiation. Thus, this may necessitate that a first batch is sent beforehand for the radiosensitivity test.
5. The cultures should be carefully sealed and packed upright (*see* Note 8, Fig. 10.11).
6. Upon reception of the irradiated *in vitro* cultures, wipe jars with ethanol to surface sterilise culture.

Fig. 10.11 Sending banana cultures for irradiation to an international or regional facility. (**a**) Sealed culture box with banana shoot tips. (**b**) The seal should remain intact if arrangements have been made for irradiation without opening the culture vessel. (**c**) Culture vessels placed tightly in vertical position. (**d**) Box ready for dispatch

3.7   Generation Advancement Under Low-Cost Conditions

1. Transfer the irradiated explants into fresh multiplication medium. This stage is labelled as M_1V_0. All explants from the same variety can be bulked as they were irradiated with the same optimum dose. No additional nomenclature is adopted.
2. Maintain cultures in growth chambers under natural light and ambient temperature.
3. After 5–7 weeks (or until buds are produced), transfer the buds into fresh multiplication medium. Label the individuals in this population as M_1V_1 (refer to Sect. 1.2 and *see* Fig. 10.8). In order to increase possibility of having mutant lines that differ, isolate each bud and transfer. When buds cannot be dissociated from the original explant or when buds are compact around the explant, then split the explant longitudinally and advance the next generation.
4. Repeat this isolation of buds for two more times every 5–7 weeks until M_1V_3 generation is reached.

3.8 Regeneration of Mutant Lines

1. Transfer M_1V_3 cultures to rooting medium without growth regulator (AC0 or AA0). Low-cost medium can be used (agar/cornstarch or agar/arrowroot).
2. Maintain under growing conditions for 4–7 weeks or until the shoots develop roots (*see* Note 9).

3.9 Low-Cost Weaning

This stage concerns post-flask low-cost hardening of the *in vitro* derived plantlets.

1. Gently remove plants from jars and wash off all gel under running water.
2. Trim back yellow/dead leaves, dead roots and extra-long roots. Leave only soft green leaves and white primary roots.
3. Plug plantlets in plastic seedling trays filled with autoclaved potting media containing peat and perlite (1:1) or sterile:manure:perlite (1:1:1).
4. Keep the trays with the plantlets under partial shade (70–80%) and high humidity (70–90%). The humidity is maintained either through a misting system or by covering the trays with a plastic sheet and regularly watering the plantlets (or misting using a spray bottle).
5. After 2–3 weeks, gradually remove the polythene sheet to allow plants to adapt to ambient temperature and relative humidity. Leave the plants for further 2 weeks under same shade conditions.
6. After 3–4 weeks when the plants have reached an average of 10–15 cm, they are ready for further hardening at higher light intensities (50–40% shade) at ambient relative humidity. The plants can be either potted individually for field planting or potted in groups into a deeper, rigid potting box with a depth of at least 15 cm for immediate screening, as follows:

    (a) Hardening for field planting: Individually transfer plugged plants into planting pots (1.5–2 l) containing autoclaved potting mix (soil:manure:perlite: peat at 1:1:1:1 or soil:manure:perlite at 1:1:1). Keep plants under 50–40% shade for about 1 month and later at full sunlight or 30% shade until they reach a height of about 30 cm for field planting. Use a slow release to fertilise the plants, otherwise use a locally available complex fertiliser high in nitrogen at the rate of ½ teaspoon per pot to avoid phytotoxicity (at 5 weeks interval).

    (b) For nursery level screening of potted plants: Transfer plugged plants in large potting boats with bottom tray to collect exudates. Use same potting media as used above. The plants can be used immediately for screening using an adapted low-cost modified double-tray system and applying same practices as highlighted in this book.

4 Notes

1. For ease of preparation of the modified MS media from the salts (Annex 10.1), it is advisable to prepare separately stock solutions of macronutrients, micronutrients, vitamins and iron salts.
2. The ½ strength of macronutrients save on mineral salt.
3. Conventionally 2 min swirling with 95% absolute alcohol can be done if available.
4. During manipulations, usual sterile methods of working near flame and regular flaming of dissecting tools after dipping in alcohol.
5. To save on PGR, use lower concentration of BAP (2 mg/l) and transfer to required BAP level once no contamination is observed.
6. Multiplication rate will depend on conditions under which plants are growing.
7. In order to prevent entry of diseases some countries, like Mauritius, have strict importation guidelines. Thus, in such cases the vessels are dispatched, irradiated and returned without opening the jar (*see* Fig. 10.11). Where the quarantine conditions for banana is less strict, the cultures can be sent in any type of container for transfer into appropriate vessels for irradiation.
   If irradiation is done locally then the explants are gently placed horizontal on the media, so as maximise chance of irradiation. Sucrose can be excluded.

   (a) For long distance (or international) transfers, the explants need to grow, and they can be gently pushed into the medium (at an angle), so that the latter receive nutrients for growth during transit.
   (b) This aspect is handled depending on quarantine exigencies of the country requiring the service (*see* Table 10.4).

Table 10.4 Conditions associated with outsourcing for mutagenic treatments

Exporting country quarantine requirement	Outsourcing for irradiation	Media	Culture vessel	Type of mutagenic treatments
The vessels should not be opened during irradiation and explants should not be taken out of vessel	Local irradiation facility	Solid, without sugar	Polypropylene (PP) or glass jar with PP cover	Only physical mutagens (X or gamma rays) can be used
	For international irradiation	Solid, with sugar	Jar to be able to fit into the irradiator Culture vessel to be quarantine sealed	
The vessels can be opened and explants can be handled during irradiation	Local irradiation facility	Solid, without sugar	PP or glass jar with PP cover or metal cover.	Both chemical and physical mutagenic treatments are possible
	For international irradiation	Solid, with sugar	Before irradiation, explants can be transferred to new vessel with no metal cover.	

8. Prior to dispatch all necessary plant import/export formalities for international transaction need to be finalized. The exporting country need to provide necessary phytosanitary certificates that satisfy the plant import conditions of country receiving the consignment. An agreement has often to be finalized to ensure that irradiated cultures are received with minimum delay.
9. As from this point the mutant line is ready for either field screening (for agronomic performance and response to field biotic stresses) or simple greenhouse screening for FW disease.

Acknowledgements The author would like to acknowledge the Food and Agricultural Research and Extension Institute (FAREI) for technical support provided. Financial assistance was provided by the Mauritius Research and Innovation Council, the International Atomic Energy Agency (Joint FAO/IAEA) and FAREI is acknowledged. Author is also grateful for the support of Mrs. Hemlata Kallydin-Gaur and Mrs. Dheema Burthia for conducting all the laboratory activities.

Annexure

Annex 10.1 Composition per litre of the modified Murashige and Skoog (1962) and preparation of stock solution

	Salt	Amount for 1 l MS medium (mg/l)	Amount of per litre of stock solution (g/l)	Voume to be used to make 1 l of medium
Stock I: Macro (10X)	NH_4NO_3	1650	16.5	Use 100 ml
	KNO_3	1900	19	
	$CaCl_2.2H_2O$	440	4.4	
	$MgSO_4.7H_2O$	370	3.7	
	KH_2PO_4	170	1.7	
Stock II: Micro (1000x)	H_3BO_3	6.2	6.2	Use 1 ml
	$MnSO_4.4H_2O$	22.3	22.3	
	$ZnSO_4.4H_2O$	8.6	8.6	
	KI	0.83	0.8	
	$Na_2MoO_4.2H_2O$	0.25	0.3	
	$CuSO_4.5H_2O$	0.025	0.03	
	$CoCl_2.6H_2O$	0.024	0.03	
Stock III: (iron salts) (200X)	$FeSO4.7H2O$	27.8	5.6	Use 5 ml
	$Na2EDTA.2H2O$	37.3	7.5	
Stock IV: (vitamins) (200X)	Myo-inositol	2.0	20	Use 5 ml
	Nicotinic acid	0.5	0.1	
	Pyridoxine HCl	0.5	0.1	
	Thiamine HCl	0.1	0.02	
	Glycine	2	0.4	

References

Ahloowalia BS, Prakash J (2004) Physical components of tissue culture technology. In: International Atomic Energy Agency (ed.) Low cost options for tissue culture technology in developing countries. Proceedings of a technical meeting, 26–30 August 2002, Vienna, Austria, IAEA-TECDOC-1384, pp 17–28

Ahloowalia BS, Savangikar VA (2004) Low cost options for energy and labour. In: International Atomic Energy Agency (ed.) Low cost options for tissue culture technology in developing countries. Proceedings of a technical meeting, 26–30 August 2002, Vienna, Austria, IAEA-TECDOC-1384, pp 41–46

Datta SK, Debasis C, Janakiram T (2017) Low cost tissue culture: an overview. J Plant Sci Res 33(2):181–199

De Guzman EV, Ubalde EM, Del Rosario AG (1976) Banana and coconut in vitro cultures for induced mutation studies. In: Improvement of vegetatively propagated plant and tree crops through induced mutations. Wageningen, IAEA, pp 33–54

Dussoruth B, Kallydin H, Burthia (unpublished) Low-cost alternatives for tissue culture

FAO/IAEA (1995) *In vitro* mutation breeding of bananas and plantains. Final reports of an FAO/IAEA Co-ordinated Research Programme organized by the Joint FAO/IAEA Division of nuclear techniques in food and agriculture (1988–1993). IAEA-TECDOC-800. https://www.osti.gov/etdeweb/servlets/purl/98209. Last accessed 5 May 2020

FAO/IAEA (2004) Low cost options for tissue culture technology in developing countries. Proceedings of a technical meeting, 26–30 August 2002, Vienna, Austria, IAEA-TECDOC-1384. https://www.iaea.org

FAO/IAEA (2011) A better banana-IAEA research helps produce higher-yielding robust variety by a. Durczok. IAEA Bull 52(2) https://www.iaea.org/newscenter/news/better-banana. Last update: 27 Jul 2017

Gitonga NM, Ombori O, Murithi KSD, Ngugi M (2010) Low technology tissue culture materials for initiation and multiplication of banana plants. Afr Crop Sci J 18(4):243–251

Jain SM, Till B, Suprasanna P, Roux N (2011) Mutations and cultivar development of banana. In: Pillay M, Tenkouano A (eds) Banana breeding: progress and challenges. Taylor and Francis, CRC Press, New York, pp 203–217, pp. 363. https://www.crcpress.com/Banana-Breeding-Progress-and-Challenges/Pillay-Tenkouano/p/book/9781439800171

Jenny C, Tomekpé K, Bakry F, Escalant JV (2002) Conventional breeding of bananas. In: Jacome P, Lepoivre D, Ortiz MR, Romero R, Escalant JV (eds) *Mycosphaerella* leaf spot diseases of bananas: present status and outlook. Proceedings of the 2nd International workshop on *Mycosphaerella* leaf spot diseases held in San José, Costa Rica, 20–23 May 2002

Jhurree-Dussoruth B, Kallydin H (2011) Investigation into low-cost medium for hardening of *in vitro* banana plantlets to promote adoption of disease-free plants. Acta Hortic 897:489–490

Kodym A, Hollenthoner S, Zapata Aris FJ (2001) Cost reduction in the micropropagation of banana by using tubular skylights as source for natural lighting. In Vitro Cell Dev Biol Plant 37(2): 237–242

Kodym A, Zapata-Arias FJ (2001) Low-cost alternatives for the micropropagation of banana. Plant Cell Tissue Organ Cult 66:67–71

Lopez J, Rayas A, Santos A, Medero V, Beovides Y, Basail M (2017) Mutation induction using gamma irradiation and embryogenic cell suspensions in plantain (Musa spp.). In: Jankowicz-Cieslak J, Tai TH, Kumlehn J, Till B (eds) Biotechnologies for plant mutation breeding, pp 55–71. https://doi.org/10.1007/978-3-319-45021-6-4

Mak C, Ho YW, Tan YP, Ibrahim R (1996) Novaria-a new banana mutant induced by gamma irradiation. In: Report of the first FAO/IAEA research co-ordination meeting on cellular biology and biotechnology mutation techniques for creation of new useful banana genotypes, 20–24 Nov.,1995, Vienna, Austria (IAEA 312. D2. R. C. 579)

Murashige T, Skoog F (1962) A revised medium for rapid growth and bio-assays with tobacco tissue cultures. Physiol Plant 15:473–497

Ogero KO, Mburugu GN, Mwangi M, Ombori O, Ngugi M (2012) *In vitro* micropropagation of cassava through low cost tissue culture. Asian J Agric Sci 4(3):205–209

Ortiz R (2013) Conventional banana and plantain breeding. ISHS Acta Horticulturae 986: VII International symposium on banana: ISHS-Promusa symposium on bananas and plantains: towards sustainable global production and improved use. https://doi.org/10.17660/ActaHortic.2013.986.19

Prakash S (1993) Production of ginger and turmeric through tissue culture methods and investigations into making tissue culture propagation less expensive. PhD thesis, Bangalore, India

Prakash S, Hoque MI, Brinks T (2004) Culture media and containers. In: International Atomic Energy Agency (ed.): Low cost options for tissue culture technology in developing countries. Proceedings of a technical meeting, 26–30 August 2002, Vienna, Austria, IAEA-TECDOC-1384, pp 29–40

Puchooa D, Purseramen PN, Rujbally BR (1999) Effects of medium support and gelling agent in the tissue culture of tobacco (*Nicotiana tabacum*). Sci Technol Res J 3:129–144

Roux NS (2004) Mutation induction in Musa-review. In: Jain SM, Swennen R (eds) Banana improvement: cellular, molecular biology, and induced mutations. Science Publishers, Enfield, pp 23–32

Sahu J, Sahu RK (2013) A review on low cost methods for *in vitro* micropropagation of plant through tissue culture technique. UK J Pharm Biosci 1(1):38–41. http://www.ukjpb.com/pdf/UKJPB07.pdf

Stadler LJ (1928) Genetic effects of X rays in maize. Proc Natl Acad Sci U S A 14(1):69–75. https://doi.org/10.1073/pnas.14.1.69

Vuylsteke DR (1998) Shoot-tip culture for the propagation, conservation and distribution of Musa Germplasm. International Institute of Tropical Agriculture, Ibadan, Nigeria, 82 pp.

Wilson V, Tenkouano A (2020) Cassava starch as alternative low-cost gelling agent for *in vitro* micropropagation of three Musa genotypes. Asian J Plant Sci Res 10(1):8–14

Chapter 11
Protocol for Mass Propagation of Plants Using a Low-Cost Bioreactor

Affrida Abu Hassan, Norazlina Noordin, Zaiton Ahmad, Mustapha Akil, Faiz Ahmad, and Rusli Ibrahim

Abstract Conventional *in vitro* mass propagation methods are labour-intensive, costly and have a low degree of automation. Bioreactor or automated growth vessel systems using liquid media were developed to overcome these problems. The use of liquid instead of solid culture medium for plant micropropagation offers better access to medium components and scalability through automation. However, the cost of setting up a bioreactor system is one of its disadvantages as such systems are expensive with limited number of manufacturers. A low-cost bioreactor system was set up using recycled, low biodegradable plastic bottles. This low-cost bioreactor, based on temporary immersion principle, has proven to be effective as a vessel for rapid plant propagation. It is designed to reduce the production cost of plant micropropagation. This chapter explains the step-by-step methods for setting up a low-cost bioreactor for banana seedling production. This low-cost bioreactor system has the potential to be adapted for large scale *in vitro* cultivation of the plant seedlings.

Keywords Low-cost bioreactor · Temporary immersion · Plant micropropagation · Seedling production

1 Introduction

The conventional micropropagation technique requires regular sub-culturing, manual handling at various stages of the process (labour-intensive) and more shelf space that contributes to high running and labour cost. Scaled-up and automated systems are therefore desirable to reduce the amount of handling, increase multiplication rates, hence overcome and/or minimize production costs of the conventional

A. Abu Hassan (✉) · N. Noordin · Z. Ahmad · M. Akil · F. Ahmad · R. Ibrahim
Agrotechnology and Biosciences Division, Malaysian Nuclear Agency, Ministry of Science, Technology and Innovation (MOSTI), Kajang, Selangor, Malaysia
e-mail: affrida@nuclearmalaysia.gov.my

© The Author(s) 2022
J. Jankowicz-Cieslak, I. L. Ingelbrecht (eds.), *Efficient Screening Techniques to Identify Mutants with TR4 Resistance in Banana*,
https://doi.org/10.1007/978-3-662-64915-2_11

micropropagation technique as initially reported by Aitken-Christie et al. (1995). This can be achieved by using a bioreactor to scale up propagation. Bioreactors are usually described in a biochemical context as a self-contained, sterile environment which incorporates liquid nutrient or liquid/air inflow and outflow systems, designed for intensive culture and affording maximal opportunity for monitoring and control over micro environmental conditions (agitation, aeration, temperature, dissolved oxygen, pH etc). The use of bioreactors in controlled condition increases the multiplication rate and plant quality and has been proven as an efficient tool for rapid production of plant cells, tissue or organ culture and metabolites. The first reported use of bioreactor for micropropagation was in 1981 for Begonia propagation (Takayama and Misawa 1981). Since then, it has been widely used and proved applicable to many plant species including cassava (Golle et al. 2019), carnation (Marzieh et al. 2017), gerbera (Frómeta et al. 2017).

Various types of bioreactor systems, with different types and different sizes of vessels and agitation mechanisms (non-agitated, mechanical or pneumatically agitated) have been developed and used as described by Paek et al. (2005), Eibl et al. (2018) and Alireza et al. (2019). Among them, temporary immersion system (TIS) bioreactor is highly suitable for use in semi-automated micropropagation. This principle of temporary immersion was first tested by Harris et al. (1983) through alternate exposure and submergence of explants by tilting a flat-bottomed vessel to opposite direction using semi-automatic system. TIS bioreactor allows immersion of explants in medium for a specific duration at specified intervals, control of contamination, adequate nutrient and oxygen supply and mixing, relatively infrequent subculturing, ease of medium changes and limited shear damage. The temporary immersion of the plant with the media is a good technique to avoid damage, since long exposure can lead to physiological malformation which causes poor regeneration. In comparison with both, solid and liquid culture systems, TIS has technological and quantitative advantages such as higher multiplication rate and reduction of production cost (Etienne and Berthouly 2002). The use of TIS for large scale micropropagation produces better plant quality and higher multiplication rate (Ziv 2005). Examples of TIS bioreactor available today include BIT® twin-flasks system (Escolana et al. 1999), Reactor with Automatized Temporary Immersion (RITA®) (Alvard et al. 1993) and Bioreactor of Immersion by Bubbles (BIB®) (Soccol et al. 2008).

1.1 Low-Cost Bioreactor System

Many established TIS bioreactor systems were patented and are quite costly, hence less preferable for large scale mass propagation. Option for a simpler and cheaper TIS bioreactor system was explored through the development of a TIS bioreactor prototype called BIO-TIS (Ibrahim 2017). BIO-TIS consists of two glass vessels, one for the *in vitro* shoots and the other for liquid culture media which is connected by silicone tubing that permits the flow of the liquid medium from one vessel to the other. It has been tested for mass propagation of horticultural crops such as: fruit trees (pineapple, banana), ornamental plants (orchids, chrysanthemums) and herbal plants (*Eurycoma longifolia* Jack, *Labisia pumila* and *Stevia rebaudiana*). In a study on pineapple propagation, the multiplication rate with BIO-TIS was found to be much higher in comparison to the established RITA® bioreactor (Ibrahim 2017).

Modification was done by replacing the glass bottles in BIO-TIS with recycled plastic bottles as an alternative for a cheaper setting up cost. A silicone cap with stainless steel tubing is fabricated for liquid nutrient or liquid/air inflow and outflow. This low-cost bioreactor is capable of supplying planting materials in large quantities for various plants, able to increase the multiplication rate of *in vitro* plantlets up to ten-fold (Ibrahim 2017; Mustapha et al. 2017), improve the quality of tissue culture plantlets by reducing vitrification and is environmentally friendly. This system can be used by the plant biotechnology industry and agro-industry to save on the production cost. Furthermore, recycling of plastic bottles helps to reduce issue of the disposal of unused material in landfills thus reduce environmental pollution from disposal of used plastic bottles.

2 Materials

2.1 Liquid Media

1. Murashige and Skoog medium including vitamins (Murashige and Skoog 1962) (Cat Nr. M0222, Duchefa, Haarlem, The Netherlands).
2. 6-Benzylaminopurine (BAP) stock solution (Concentration: 1 g/1 l) (Cat Nr. B0904, Duchefa, Haarlem, The Netherlands) (*see* Note 1).
3. α-Naphthalene acetic acid (NAA) stock solution (Concentration: 1 g/1 l) (Cat Nr. N0903, Duchefa, Haarlem, The Netherlands) (*see* Note 2).
4. Sucrose (table sugar).
5. Sodium hydroxide (NaOH).
6. Hydrochloric Acid (HCl).
7. Distilled water.
8. Beaker (Volume: 1 l).
9. Bottle (Scott, Volume: 1 l).
10. Spatula.
11. Pipette.
12. Pipette tips.
13. pH meter.
14. Electronic balance.
15. Autoclave.

2.2 Bioreactor System *(Fig. 11.1)*

1. Recycled plastic bottle (Capacity: 5-10 L).
2. Silicone cap.
3. Silicone tube (Diameter: 6 mm).
4. Air filter (Sartorius, Midisart® 2000 PTFE or equivalent).
5. Air compressor pump (2 unit) (Rocker 320, Taiwan or equivalent model).
6. Timer (2 unit).
7. PVC Pipe (Length: 1 m, Diameter: 15 mm).
8. Cling film (3 cm width).
9. Scissors.

2.3 Culture Initiation

1. 30–40 day old *in vitro* shoots of banana measuring 1.5 cm with 2–4 leaves.
2. Forceps.

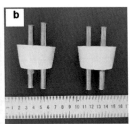

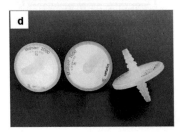
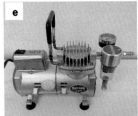

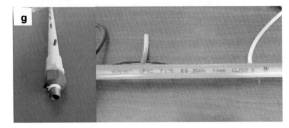

Fig. 11.1 (**a**) Recycled plastic bottles. (**b**) Modified silicone cap. (**c**) Silicone tube. (**d**) Air filter. (**e**) Air compressor pump. (**f**) Timer. (**g**) PVC pipe

3. Scalpels.
4. Blades.
5. Ethanol (70% v/v).
6. Tissue paper.
7. Hot bead sterilizer or bunsen burner.
8. Laminar air flow cabinet.

3   Methods

3.1   Preparation of MS Medium

1. Fill 800 ml of distilled water into a beaker.
2. Add MS medium, 30 g of sucrose, 2.5 ml of BAP stock solution and 0.1 ml NAA stock solution.
3. Stir the solution until dissolved.
4. Adjust the pH to 5.6–5.8 by adding NaOH/HCl.
5. When the desired pH is achieved, bring the volume to 1 l with distilled water.
6. Transfer the media into a bottle and sterilize for 15 minutes at 15 p.s.i and 121 °C.
7. Allow medium to cool down to room temperature prior to use.

3.2   Preparation of Bioreactor Set and Culture Initiation (See Note 3)

1. Sterilize plastic bottle using 5.25% sodium hypochlorite solution (*see* Note 4).
2. Prepare the modified silicone cap by connecting tube, cap and air filter as shown in Fig. 11.2 and sterilize by autoclaving at 15 minutes/15 p.s.i/121 °C.
3. Inside a laminar air flow, take two sterile plastic bottles of the same size (*see* Note 5).
4. Fill 2 l of MS media into the first bottle and close the bottle with the sterile modified cap set.
5. In another bottle, transfer approximately 20 gram of *in vitro* shoots and close the bottle with the sterile silicone cap set (Fig. 11.3).
6. Wrap around the cover using cling film to ensure that no leakage occurs during operation.
7. Connect both bottles with silicon tubing as shown in Fig. 11.4.

3.3   Setting up Low-Cost Bioreactor

1. Connect the low-cost bioreactor set to the air compressor pump (Fig. 11.4). A number of low-cost bioreactors can be combined and run simultaneously as

Fig. 11.2 A close view of the modified silicone cap that consists of a fabricated silicone cap, stainless steel tubes to allow media/air outlet and inlet, and silicone tube

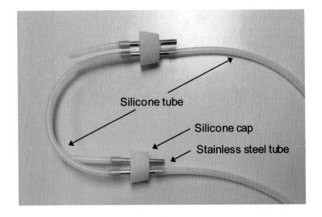

Fig. 11.3 Explants are transferred into sterile plastic bottle under aseptic condition in the laminar air flow cabinet (left). A complete low-cost bioreactor ready for incubation (right)

shown in Fig. 11.5 and Fig. 11.6. For this protocol, the system was tested for a max of 4 sets merged together. However, the efficiency of the system must be revaluated if there is additional set added.

2. Program the timer of Air Compressor Pump 1 for 15 minutes to ensure all the media is transferred into the explants-containing bottles.

3. Immerse the *in vitro* culture for 30 minutes.

4. Program the timer of Air Compressor Pump 2 for 15 minutes in order to transfer the media back into the first bottle after the immersion is completed.

5. The immersion cycle for this system is every 6 hours. The immersion cycle is further explained in Fig. 11.7.

6. Place the bioreactor system in incubation room with 25 ± 2 °C and 16 hours photoperiod for a period of two months (*see* Note 6).

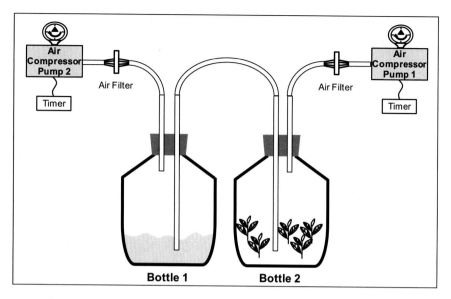

Fig. 11.4 Two bottles of the same size are used. Bottle 1 is filled with media and Bottle 2 is for explants

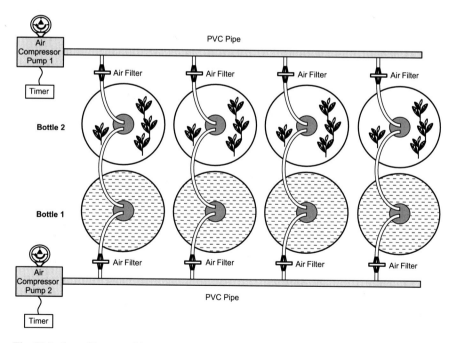

Fig. 11.5 Sets of low-cost bioreactor can be combined and run simultaneously

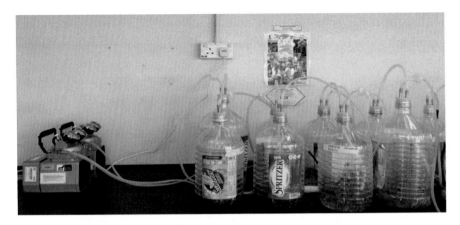

Fig. 11.6 Large-scale production of banana plantlets through low-cost bioreactor

3.4 Harvesting

1. Detach all silicone tube from the cap and take out the cap from the culture bottle.
2. Carefully tilt the bottle and take out the plantlets using forcep.
3. Rinse the plantlets under running tap water before hardening.

4 Notes

1. To prepare 100 ml of BAP stock solution (1 g/l), dissolve 100 mg of BAP with 2–5 ml of 1 N NaOH. Then bring the volume to 100 ml with distilled water and mix well. The stock solution can be stored at 4 °C for several months.
2. To prepare 10 ml of NAA stock solution (1 g/l), dissolve 10 mg of NAA with 1–2 ml 1 N NaOH. Bring the volume to 10 ml using distilled water and mix well. The stock solution can be stored at 4 °C up to several months.
3. This protocol has been optimised for propagation of banana. However, the same protocol can be applied for propagation of other plants using a suitable media and optimisation should be carried out to make sure that the system is suitable.
4. Sterilisation of plastic bottles is performed using 5.25% sodium hypochlorite solution. Pour 200 ml sodium hypochlorite solution into bottle and shake for 5 minutes. Repeat this step for another 2 times. Lastly, rinse with sterile distilled water twice to ensure no trace of sodium hypochlorite left. This sterilisation work is done in the air laminar flow cabinet. Alternatively, gamma irradiation can be used to sterilise the plastic bottle.
5. Preparation of plastic bottle and *in vitro* initiation work must be performed using aseptic techniques in the laminar air flow cabinet to avoid contamination.

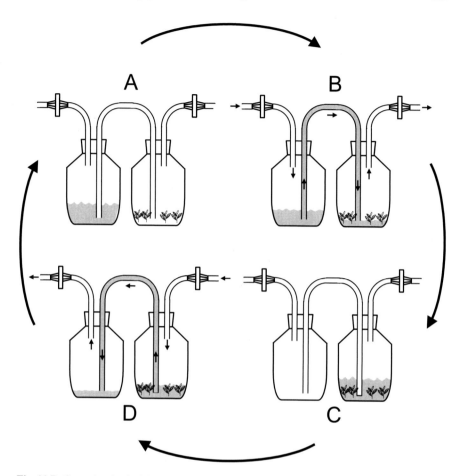

Fig. 11.7 Operational principle of low-cost bioreactor system: (A) The liquid medium is located in Bottle 1 and explants in Bottle 2 (B) The air compressor pump 1 is run for 15 min and all medium flow from Bottle 1 to Bottle 2 (C) The explants are immersed into the liquid medium for 30 minutes (D) After immersion is complete, the air compressor pump 2 is run for 15 min to allow all medium flow back into Bottle 1

6. Due to the efficient gaseous exchange between plant tissue and gas phase inside the vessel, the yield obtained from this procedure can increase up to 10–15 old compared to solid media.

Acknowledgments Authors wish to thank Ms. Norhayati Irwan, Ms. Nashimatul Adadiah Yahya and Mr. Nor Hafiz Talib for their dedication and assistance. We would also like to thank Malaysian Nuclear Agency for their continuous support. This work was funded by the Ministry of Science, Technology and Innovation (MOSTI) of Malaysia under MOSTI Social Innovation Funding.

References

Aitken-Christie J, Kozai T, Takayama S (1995) Automation in plant tissue culture - general introduction and overview. In: Aitken-Christie J, Kozai T, Smith MAL (eds) Automation and environmental control in plant tissue culture. Springer, Dordrecht, pp 1–18

Alireza V et al (2019) Bioreactor-based advances in plant tissue and cell culture: challenges and prospects. Critical Rev Biotech 39(1):20–34

Alvard D, Cote F, Teisson C (1993) Comparison of methods of liquid medium culture for banana micropropagation. Plant Cell Tissue Organ Cult 32(1):55–60

Businge E et al (2017) Evaluation of a new temporary immersion bioreactor system for micropropagation of cultivars of eucalyptus, birch and fir. Forests 8(6):196

Eibl R et al (2018) Plant cell culture technology in the cosmetics and food industries: current state and future trends. App Microb Biotech 102:8661–8675

Escalona M, Lorenzo JC, González B et al (1999) Pineapple (*Ananas comosus* L. Merr) micropropagation in temporary immersion systems. Plant Cell Rep 18:743–748

Etienne H, Berthouly M (2002) Temporary immersion systems in plant micropropagation. Plant Cell Tissue Organ Cult 69(3):215–231

Frómeta OM et al (2017) *In vitro* propagation of *Gerbera jamesonii* Bolus ex Hooker f. in a temporary immersion bioreactor. Plant Cell Tissue Organ Cult 129(3):543–551

Golle DP et al (2019) Temporary immersion bioreactors: establishment of cassava. J Agric Sci 11(4):176–181

Harris E, Edwin B, Mason B (1983) Two machines for in vitro propagation of plants in liquid media. Can J Plant Sci 63(1):311–316

Ibrahim R (2017) The potential of bioreactor technology for large-scale plant micropropagation. Acta Hortic 1155:573–583. https://doi.org/10.17660/ActaHortic.2017.1155.84

Marzieh A et al (2017) Micropropagation of carnation (*Dianthus caryophyllus* L.) in liquid medium by temporary immersion bioreactor in comparison with solid culture. J Gen Eng and Biotech 15(2):309–315

Mustapha A, Affrida AH, Norazlina N et al (2017) Low-cost bioreactor for rapid, mass and cheap production of plant seedlings. Paper presented at Nuclear Technical Convention 2017, Bangi, Malaysia, 13–15 Nov 2017

Paek KY, Chakrabarty D, Hahn EJ (2005) Application of bioreactor systems for large scale production of horticultural and medicinal plants. Plant Cell Tissue Organ Cult 81:287–300

Soccol CR, Scheidt GN, Mohan R (2008) Biorreator do tipo imersão por bolhas para as técnicas de micropropagação vegetal. Universidade Federal do Paraná Patent (DEPR 01508000078)

Takayama S, Misawa M (1981) Mass propagation of *Begonia* × *hiemalis* plantlets by shake culture. Plant Cell Physiol 22(3):461–467

Ziv M (2005) Simple bioreactors for mass propagation of plants. Plant Cell Tissue Organ Cult 81:277–285

Printed in the United States
by Baker & Taylor Publisher Services